THIS ORDINARY STARDUST

THIS ORDINARY STARDUST

A SCIENTIST'S PATH *from* GRIEF *to* WONDER

Alan Townsend

NEW YORK BOSTON

Cover design by Dana Li
Cover photo of landscape by Cavan Images-Offset/Shutterstock
Cover photo of outer space by Shutterstock

Grand Central Publishing
Hachette Book Group
1290 Avenue of the Americas, New York, NY 10104
grandcentralpublishing.com
twitter.com/grandcentralpub

First edition: June 2024

Grand Central Publishing is a division of Hachette Book Group, Inc. The Grand Central Publishing name and logo is a trademark of Hachette Book Group, Inc.

Print book interior design by Bart Dawson.

Library of Congress Cataloging-in-Publication Data

Names: Townsend, Alan R., author.
Title: This ordinary stardust : a scientist's path from grief to wonder / Alan R. Townsend.
Description: First edition. | New York : Grand Central Publishing, 2024.
Identifiers: LCCN 2023041436 | ISBN 9781538741184 (hardcover) | ISBN 9781538741207 (ebook)
Subjects: LCSH: Townsend, Alan R. | Townsend, Alan R.—Family. | Spouses of cancer patients—United States—Biography | Brain—Cancer—Patients—Family relationships. | Religion and science. | Biogeochemistry. | Grief.
Classification: LCC RC280.B7 T67 2024 | DDC 616.99/4810092—dc23/eng/20231114
LC record available at https://lccn.loc.gov/2023041436

ISBNs: 9781538741184 (hardcover), 9781538741207 (ebook)

Printed in the United States of America

LSC-C

Printing 1, 2024

True solace is finding none, which is to say, it is everywhere.

—Gretel Ehrlich,
The Solace of Open Spaces

PROLOGUE

Decades ago, I became entranced by the stardust that resides in all of us. I sat in a California lecture hall and listened to my biology professor talk about the ways we pass that stardust not only between one another but to everything else on earth. About how those exchanges can happen in the tiniest of spaces or across the world, sometimes every second, sometimes not for millions of years. And about how our own species was rewriting the rules of the game.

It hooked me. And while the whole "we are made of stardust" thing is also a total cliché, when the lecture ended, I thought: that stupid hippie bumper sticker was a true and literal statement of who we are, our nature as well as our limits. When viewed in our most elemental form, people are trillions of outer-space atoms, moving around temporarily as one, sensing and seeing and falling in love. Then those atoms scatter, joining one new team for a bit, then another. Far from depressing, I found this notion profoundly comforting. Sure, "I" the atomic collective wouldn't be around all that long in the grand scheme of things—but my atoms would, forever remixing and reencountering one another. I thought: No matter what happens, we're still here. And we always will be.

Our never-ending story began at the dawn of time, when the formation of the universe blew out clouds of hydrogen and helium. For a while that's all we had, barring a few traces of lithium and beryllium. Then things cooled off a bit as stars were born and coalesced into galaxies. The heat inside those stars became cosmic ovens in which the lightest elements fused into progressively heavier ones, including the stuff of life. Carbon, nitrogen, oxygen. Along with hydrogen, these elements are most of me, you, your dog, and your houseplants.

But it doesn't end there. Like a meal without spices, we are nothing without twenty or so other elemental building blocks, arranged in myriad combinations that create the mind-boggling array of life on this planet. Forging the raw material to build that life took the heat of those long-distant stars, including some truly blistering ones that would put our own sun to shame. To get oxygen or anything heavier, atomically speaking, you need a stellar oven that can dial the heat up past one billion degrees.

But even an unimaginably hot furnace doesn't create a very interesting world on its own. Life is what makes the whole recipe transform into something unexpected and miraculous. Living things, simply by going about their daily business and then passing on, exchange the planet's stardust not only among one another but into rocks and water and air, seashells and coal and carbon exhaust. As we have for billions of years, we the living send stardust out, take it back, hold it for a while, and send it out again. A constant rhythm of movement, rot, and growth marks every minute, hour, year, and generation. One by-product of all that movement? In your lifetime, you might contain some part of every human ever.

I dove into the study of these elements, and the deeper I went, the more fascinating it became. The world was connected in ways I never dreamed. For example, forests near my Hawaiian birthplace depend on elements contained in dust that blows all the way from the deserts of Mongolia. They also rely on the ocean below them, for other elements that move from sea to air before raining back down to bathe their roots. And the distribution of the elements in the soils below those trees could reveal secrets about how past Hawaiian societies had lived and perhaps why they succeeded or failed.

The hidden secrets of life's elements helped me appreciate how every one of us, and every living thing, exists because of communities each life form helps to construct and sustain, and also because of ones we will never know, some of which flourished and then disappeared a very long time ago. Humans carry around an uncountable number of bacteria that constantly take up a few of our elements, give them back, and in so doing help us digest our food, ward off disease, and generally make it through the day. In the middle of the Amazon rain forest, there are trees that may owe their existence to species of lumbering animals that are now extinct but who once would carry nutrients from the rich soils of the lowland riverbanks to the more impoverished interior, depositing them into the soil in their excretions and ultimately from their decaying bodies. Those same trees are also literally coated in bacteria and fungi, many of which help them access the very elements they need to survive.

I began to realize how tracing the elements of our lives shows us how we are all just moments in time, but ones that reflect a richness of stories that precede our existence and others that will continue well after we are gone. The study of these elements—the science I have pursued for decades now—is known as biogeochemistry. The word is overly stuffed with prefixes because it is part biology, part geology, part chemistry, part a whole bunch of other things too. At times physicists joke that every other branch of science is a subsidiary of their own, but biogeochemists have a good hand to play here. Given long enough, the elements that assemble life can pass from pretty much anything and anywhere to anything and anywhere else. They connect us to their galactic birth at the beginning of time and to the entirety of the world. Mary Oliver put it beautifully in "Sister Turtle," one of the essays in her book *Upstream*:

> All things are meltable, and replaceable. Not at this moment, but soon enough, we are lambs and we are leaves, and we are stars, and the shining, mysterious pond water itself.

The science of biogeochemistry has much to teach us. Why this cornfield needs more fertilizer than that one. Why too much of that fertilizer will kill sea life hundreds of miles away. Why one lake looks pea green and another cobalt blue. Why our ancestors flourished in one place and not in another. Even why our planet is warming.

Biogeochemistry also shows us the inevitability of life's ebbs and flows and that not all swings are the same. Sometimes, the rules of the game are changed in a hurry, and everything and everyone must scramble to adjust. Oxygen fills our atmosphere. A massive volcano eruption or meteor strike blocks the sun. Plants evolve and spread around the world. Humans arrive and largely just hang around for millennia, then transform the entire planet in only a few generations.

But it took my world being ripped apart for me to realize one of the most important lessons of my science, of any science: it can nurture the soul.

That lesson didn't come quickly. There was no lecture hall *aha* moment. For years I believed that science held the keys to nearly every answer humanity sought but that those answers were about solutions—technical, medical, environmental—not about a way of being in the world. I was sure that science could, with sufficient knowledge, explain our planet and ourselves and predict so much of what might come next. And I thought that was its essential purpose: no more and no less.

To be fair, what science *can* predict, answer, or invent is astonishing. But an evangelistic belief in science as the source of all answers is both limiting and dangerous. In part, that's because science is a field created and practiced by human beings, which means it's often a mess. Its miracles sprout at less-than-predictable times and from an ample crop of failure, and that's just the beginning of its problems. Science is replete with histories of oppression, abuse, exclusion, and violence.

Yes... and here's what I sometimes tell students who are embarking on a scientific path: this wild, chaotic, painful, loamy field is also the site of humanity's greatest potential, whether we're talking about each of us as individuals or all of us together as humankind. It's where we can turn our love for one another into tangible and practical service. Science invites us to sow all that is best about us—curiosity, caring, altruism—and in time, with luck, reap the kind of innovation that saves and nourishes not just our loved ones but millions of other people too.

I believe this because I was reminded in the most agonizing way possible what it really means to be a scientist. It's not an achievement, a role, or a superior understanding; it's a process, a way of observing and existing in the world. It isn't about being any less pathetic or mortal than anyone else; it's about getting away from our own ego and learning to wonder—no matter what hits may come—in the teeming, crushing, profound, gorgeous, surprising, and extraordinary world that *is*.

Because in time I came to see that far from antithetical to faith or spirituality, science offers hope that life on earth can make its way through the eye of any needle, that our individual choices matter, and that love can bring us back from the brink of annihilation. A scientist's mindset doesn't just involve finding cures and creating new technologies; it can be, if we let it, a practice of spiritual self-salvation. An act of love.

I'm not a Christian. I don't go to church. But as I've learned—sometimes painfully—science and religion have far more common ground than I once thought. If, as the biblical Paul wrote to the Corinthians eons ago, love is patient and kind, insisting not on its own way but rejoicing in the truth, then perhaps there is no purer form of love than science. It is an act of profound attention and empathy. It requires a willingness to come back and back and back to a problem, no matter how numerous or embarrassing your failures, in the quest for a solution. Science is the discipline of figuring out the things one can change and learning how to work with the things one can't. On so many levels, these are lifesaving skills.

I didn't see it this way, not yet, when seven simple words changed my life:

We found something on your daughter's MRI.

CHAPTER ONE

Nearly twenty years before, I was mostly concerned with my right heel, because every step I took through the broiling Amazonian pastures was accented with a stab of pain. That night I sat on the edge of a decomposing porch, remaining outside because the cabin's rough-hewn interior stunk of dirty field clothes and held the stifling tropical heat. I trained a headlamp on my heel to reveal an angry sweep of red, a white circle, and a small black dot at its center.

I turned to the man who was seated to my left. He was tall and muscled with close-cropped black hair, and his dark eyes were focused on the tip of a machete he was using to pry bits of dried mud from the sole of his boot. Greg was the reason I'd come to this part of the Amazon, after he swapped life as a Navy Special Forces officer for one as a tropical ecologist. I pointed at the black dot.

Any idea what this is?

He leaned over, then resumed the machete picking.

Bicho de pe.

What?

Bicho de pe. Foot beast. Get a knife and dig it out, or she'll swell up and start laying eggs.

Jesus, what?

But I did as he instructed. I began by draining one of the airplane bottles of Jack Daniel's we had smuggled into the dry Brazilian Forest Service camp, then opened the smaller blade on my pocketknife, burned its tip to a gunmetal blue, and poked gingerly at the foot beast. It hurt. Another ineffective stab proved too much for Greg.

Stop fucking around and dig it out already.

The camp was known only by its nearest kilometer marker on the pummeled road that led from the river city of Santarém, Brazil, into the deep heart of the central Amazon. Kilometer 83. A scattering of woodbox cabins, one of which sprouted a dented satellite dish that allowed discordant blasting of soccer games from all over the world, sometimes late into the night. The ubiquitous chickens. Four cadaverous dogs. And a tom turkey who somehow survived despite his piercing *gobblegobblefuckinggobble* around three thirty each morning. Which in turn set off the dogs. We'd lie awake, muttering obscenities and occasionally bumping against a neighbor's hammock until, by four thirty, the howler monkeys' asthmatic roars signaled the imminent dawn. At least the coffee was good.

Earlier that day, when my siphoning attempts to refill our pickup failed, Greg had started the flow from a rusting barrel mounted to the truck's bed by sucking on the hose until he spat out diesel fuel in a violent spray.

There. Now let's get the hell out of here.

Then he sent us careening down the road, our passage hazardous for the occasional wayward chicken, me trying not to have my head go through the roof of the battered white Toyota. One lucky rooster leaped before us in terror, rode the front grille for a few seconds, then somehow hit the opposite side at a dead run. It was the middle of the dry season in this part of the Amazon, and so everything was bathed in dust, including the inside of my mouth. I spilled a swig of water on my shirt when the pickup bucked yet again, while Greg laughed a bit maniacally at me and the rooster at once. He seemed to elevate the chaos on purpose, as though he needed it to feed something inside.

Now, in the dim light of the cabin's porch, I told him to *shut the hell up* as he leaned near my foot, happy to have me be the post-dinner entertainment. I shoved the knife in harder, angling beneath the dot, and gritted my teeth at the pain. Another quick push released a thin stream of opaque fluid, and there it was, slumped on the tip of the blade. A nondescript crumb of black and brown, easily mistaken for a flake of the decaying cabin. I dropped it on the ground, and Greg returned to his boot picking with an air of disappointment that the show was over. Thirty yards away, yet another truck growled past in a cloud of dust and diesel, its trailer laden with logs from the Amazonian interior.

The big-money trees were often targeted first. Mahogany. Ipê. Jatoba. Cut beside their less lucrative neighbors, dragged out by bulldozer or sectioned in place and removed by horse and chain. For every tree destined to be multi-thousand-dollar tables and cabinets, another ten were felled to make way for its removal. Then many of those left behind were hacked in a more obvious sweep until a tangled mess of logs, stumps, vines, broken branches, and smaller trees were all that remained. Most of the bigger animals were now gone, escaping to nearby forest cover, assuming it was there. Those who could not travel as quickly—snakes, frogs, lizards—scuttled around. For the moment.

When the dry season is firmly in place, a forest with countless holes now punched in it and thus barely worth the label crisps quickly in the relentless heat until the whole thing is set on fire. The stumps of trees once a hundred or more feet tall will smolder for weeks, their heat enhanced by a landscape carpeted in charcoal. If you walk through it at midday, your feet will feel hot through your boots and your breath will feel labored, as though you are forced to pass through an unending sauna. The month before we arrived, so many contrails of deforestation smoke intertwined above Santarém that planes could not fly.

When the forest burns, elements the trees once contained are left on the charred ground in a layer of ash. It's hard to imagine that potential bounty still exists in such a dystopian landscape, but when the rains come, the secret's out. A new crop—cassava perhaps, more recently, soy—or the seeds of pasture grass will sprout with abandon, fueled by the nutrients contained in the ash. In a pasture just a year or two removed from the burn, the grass can scrape the bellies of the flop-eared cattle as they push through a thick shag carpet of brilliant green. This pulse of fertility is a big reason fires are set in the first place.

But it's all an illusion, draped unsteadily on the bones of the former jungle. Because many of the nutrients once held in the towering forest and passed from one tree to another over centuries quite literally went up in smoke and blew away. The lush grass is on borrowed time, and when it begins to fade, the descending cycle continues as yet another fire is set, life-giving nutrients already in short supply in the ancient soils going up in smoke once more. Within a decade, give or take, patches of heavily trodden earth are everywhere, and the grass is low and coarse and beset with thorn-ridden brush. The cattle are now skeletal. The search for a new patch of forest to burn will begin again.

Novelist Richard Powers put it this way in *The Overstory*:

> That's the trouble with people, their root problem. Life runs alongside them, unseen. Right here, right next. Creating the soil. Cycling water. Trading in nutrients. Making weather. Building atmosphere. Feeding and curing

and sheltering more kinds of creatures than people know how to count.

He's right. Those Amazonian forests are not only a remarkable cradle of life, but they literally create their own weather. Cut them down, and the rains they seed in their daily business of piping water from soil to air starts to go away. Lose enough of the forest, and it might never return.

For weeks, we sweated buckets throughout these failing pastures. We were there to understand more about the cycle of deforestation and hopefully learn something about how to slow it down. Greg's initial path to the navy had included training as an aerospace engineer, and while he was stationed for a few months in Hawaii, he became fascinated by issues surrounding species conservation and land management. He began to wonder how his background might allow him to measure and map such problems in ways even an army of people on the ground could not.

Now he was putting those skills to use. If everything in our admittedly ambitious plans worked, we'd be able to predict which pastures might degrade quickest, which might hang on for longer than expected, and from that how to best target better preservation of the forests—all from satellites flying far above. But first that meant collecting a whole bunch of data on the less glamorous ground.

Some days we'd lay out a grid in a pasture using a fifty-meter forestry tape. Then we'd walk the lines of that grid carrying a yellow metal box about the size of a toaster oven. The box had a digital control panel on top and a series of adjustment knobs, as well as a flexible six-foot cord that emerged from one side; at the end of the cord was a handheld sensor that we would point at the ground as we walked. The device collected high-resolution data on the light we can see... and that we can't. The yellows and greens of the grasses, the earthen tones of the soil, the occasional blue or purple of a tiny flower somehow hanging on in these pastures—all of those colors are part of something known as the electromagnetic spectrum, within which only a fairly small fraction is visible light. But there's a great deal of information in the wavelengths we cannot see but that still reflect off just about any surface. And some of those wavelengths could give us information about the chemical composition of the grasses. Those data, in turn, were an indicator of how healthy—or not—a given pasture might be. And from that, we hoped to go back to the satellite data to figure out what we could piece together from them. Could we see hints of those chemical changes? Could we then map that across the entire region, and from that, make some predictions about what areas were most at risk?

Carrying the box was usually a pretty fun day. Digging soil pits in the pastures was not. The ancient soils were composed of a reddish-yellow clay that would compact under the cattle's hooves and sometimes seemed more like cement than soil. At one point as our heads swam in the heat while our shovels chipped away embarrassingly small amounts of dirt, Greg leaned on his before saying, *Maybe we should have brought a goddamn jackhammer.*

But getting that dirt was also essential to our plan. The soil was the ultimate source of nutrition to the pastures, and while the satellites could not see into the soils themselves, our hope was that information on the grasses from the handheld sensor, coupled with soil maps of the region and our own chemical analyses of those soils, would let us piece together that predictive map driven by those satellites.

It was hard work, but both of us were young and fueled by an optimism that a few scientific answers would turn the tide. We thought that if we could figure out where and why the pastures were failing, that information would both help the farmers and ranchers get more out of a given piece of land . . . and by extension, not have to cut down as much of the forest. It was, to be honest, a bit idealistic. But not totally bonkers. Years later, Greg would build on our work to create tools that were used at the highest levels of government to help understand patterns of deforestation and what was driving them. He'd even end up transforming that little handheld box into an entire airplane of his own design, one that would end up mapping large swaths of rainforest—and then, of all things, coral reefs—with a level of stunning detail that neither of us could have predicted in those early days.

Not having our own airplane yet, we had to focus our energies on a smaller scale and spent many of our days toiling in a group of sorry-looking pastures that lay just beyond a weather-beaten line of shacks and an open-air restaurant and bar. The latter faced the main road to the Amazon interior, about thirty kilometers south of our field camp. The owner of it all was a man named Manuel, who had a mat of greasy black hair and a deeply lined face that was perhaps a road map of the complicated ways he seemed to run his world. Out in the pastures, he scrambled to keep a scattering of rawboned cattle alive. In the restaurant and bar, he brought soft drinks or beers and platefuls of simple food to an uncertain and sometimes volatile mix of travelers, most of whom were truckers hauling logs or a random mix of other wares. A handful of young women helped Manuel serve the food and drinks, but in time we learned they also led some patrons to the small shacks beyond the bar. Upon their return, they'd confer quietly with the owner.

The first time we met Manuel, Greg presented him with a poster-size satellite image of the entire region, then pointed out the road, the buildings, and the patchwork of uneven squares that represented his land. Manuel put it in a central position over the bar immediately, covering up another poster that featured a barely clothed model advertising a well-known soccer brand. He called a few others over to have a look and held forth for a while in rapid-fire Portuguese, much of which we could not follow. The men around him occasionally nodded or asked a question, while Greg and I stood to the side with the goofy and uncertain smiles of those who aren't really in the loop.

One afternoon late in the trip, Manuel stood beneath the poster while once again talking quietly to one of the women, then walked over to the corner table where I was drinking a Coke, the bottle sweating in the heat. Greg was still somewhere out in the fields, calibrating the yellow box for our next day's work. Manuel beckoned me to follow him out to a different pasture, and we walked slowly in the hot sun before he leaned over to pull up a clump of dry grass. He sighed and spoke in Portuguese while I struggled to understand.

You see how pale and tough the grass is? Not good for the cows.

Then he pointed the offending blades at a thicket of thorny bushes.

And more of these each time I burn. They are bad too.

Why burn it then? I asked.

Worse if I don't. No choice.

Then his face brightened a little.

Come, let me show you something different.

He led me across the pasture toward a curving line of trees. As we drew close, I could see that they bordered a small creek, its water tannin-stained but clear. Thigh-high and now bright green fronds of the same pasture grass obscured its banks, the blades softer here, all of it beneath the welcome shade of a canopy far above. One tree had fallen at an angle across the creek, and Manuel removed his scuffed rubber boots before wading in and then sitting on the log, his feet now dangling in the water. Then he suddenly jumped up and stood in the creek, looking back at the log as he pointed at an enormous ant coming across the top of the fallen tree.

Tucandeira.

I recognized the word. Known in English as the bullet ant, the insect is notorious for landing on the top of "most painful animal bites" lists. There was even a poor (crazy?) bastard by the name of Justin Schmidt who codified all the suffering into a pain scale known as the Schmidt sting pain index. This arose out of a scientist, well, being a scientist. Schmidt created the pain scale as an offshoot of his research into the ways insect venoms can affect our blood. An entomologist, he also happened to have the misfortune of being stung or bitten by most of the creatures on his list, and the bullet ant took the top prize for both the intensity and duration of the pain its neurotoxin inflicts. It is sometimes known as the twenty-four-hour ant. Schmidt passed away recently, but he maintained his sense of humor amid all the assaults he absorbed, describing one offender that comes in just behind the bullet ant as "like being chained in the flow of an active volcano" before writing, "Why did I start this list?"

Not being a crazy entomologist, I too avoided the log, taking off my boots and wading into the creek. After the heat of the pasture, the feel of the water on my bare feet was about as far from Schmidt's index as you could get. I stood next to Manuel for a few minutes and took it all in, my mind trying to sort out his obvious love for this place with the full scope of what I now knew was going on in and behind his bar.

I pointed at a huge tree downstream. Its bark was smooth and the color of coffee with a dollop of cream, and the trunk flared at its base into a series of deep folds.

These trees are worth a lot, yes? Is this still your land?

He nodded.

Yes, but I don't cut them.

Why not?

Because then I could not come here in the mornings and watch the birds.

He paused while I tried to understand the complexity of the man. He was apparently willing to make money from roadside prostitution but not by cutting down these trees. Manuel looked at me and said a few words I'd remember for a long time. When he spoke, it was soft, and it almost seemed like he was talking to himself.

We make choices in life, yes? And they all mean something.

In the moment, he seemed resigned and self-reflective. I wanted to believe he was questioning some of his choices, but in truth, I was probably just seeking a tenuous peace with the fact we had chosen to associate with this man at all. Years later, though, as I grappled with the ways my own life was suddenly blown apart, I thought back to how Manuel was just one more example of how very little in life is simple and how his words hinted at one of the real mindfucks of human existence. We can't control the chaos of the world, yet our choices matter to that world so very much. The carbon dioxide we choose to release into the atmosphere with our everyday activity—car and plane trips, home heating, and yes, what was coming out of those burning trees Greg and I were there to study—stays there for a thousand years, writing a part of our legacy into the sky. Manuel's choices to cut some of those trees down and our collective choices to go buy the hamburgers his pastures might eventually create, all of it had consequences for those who would follow us.

Our life choices similarly write lasting change within our bodies. Choosing hope over fear results in distinctly different chemical cascades within. If we can remain hopeful in the face of life's chaos, our bodies are bathed in hormones that both settle and invigorate us; slip into fear and stress, and a different cocktail goes on the attack. Remarkably, trauma and our reactions to it affect not just our own health but can alter the gene expression—and hence the bodies—of our descendants too. Our choices really do have lasting consequences. That isn't belief; that is scientific fact.

One understandable reaction to this fact is paralysis. So often, we obsess over the consequences of a choice, even when we don't step back and realize how long a given choice may reverberate through the world. At one point on that trip, Greg and I floated in a remote jungle river bordered by soaring trees and chattering birds and even watched in wonder while a group of pink river dolphins swam by. Yet the beauty of our surroundings faded when we began to talk about our marriages and the radiating cracks in their foundations. As so often happens, stress and anger took hold as we ground through the unwelcome choices we each might have to face, until at one point Greg finally paused, looked around, and said, *Look where we are. What the fuck are we doing?*

As Manuel and I stood in the creek that day, I stumbled through my crappy Portuguese and tried to tell him more about the scientific questions we were after. He began to ask his own questions, and at one point I told him—partly in words, partly in comical moments of pantomime—about how the trees beside this creek might depend on nutrients contained in dust that blows all the way from the Sahara Desert and how some of that dust might be right inside the leaves above our heads.

When he got it, the visible stress over his struggling cattle and who knows what else evaporated, replaced by a wholly unexpected look of delight. This was followed by a peppering of questions that took us a long time to sort out. How could African dust be inside a leaf? How did that dust get here? How could I possibly know it was from across the ocean? In him, I saw how curiosity, no matter how difficult our circumstances are, not only leads to answers that *can* help, but it literally settles us down. Just like the balm that comes from hope, a curious mind eases our nerves, opens our hearts, and puts the brakes on those stress hormones that eat us from within. Curiosity—a bedrock of the scientific mindset—is medicine in itself.

CHAPTER TWO

Do you know what butterfly metamorphosis looks like? It's not like what happens to a tadpole. Spindly legs and rainbow wings don't just poke their way out of a cocooned caterpillar's body. What happens is that after caterpillars encase themselves in a chrysalis, they dissolve (or nearly dissolve; this depends on the species). If you cut their chrysalis open at this moment, all that will come out of most is a splat of runny goo: a baby bug liquefied, an old life gone forever.

The sight is horrifying. It looks so much like death that for centuries naturalists thought that's exactly what it was. Metamorphosis looked like resurrection. Theologians claimed that it was God's way of proving the Jesus thing was possible.

The truth of how metamorphosis actually works is different, yet in my mind it's equally miraculous. The liquid itself is a bunch of digestive enzymes. Within the liquid of the chrysalis, invisible to the eye, are tiny cell clusters called imaginal discs.

Using the enzymes as fuel, the imaginal cells begin to multiply, forming wings and legs and a brain and eyes, and from there the whole body of a butterfly.

Just a decade ago, Georgetown University biologist Martha Weiss discovered that some of these imaginal cells contain the caterpillar's memories, holding the wisdom of the past through the dissolution of the present and into a transformed future. The chrysalis is perhaps not, then, a living metaphor for resurrection, but for hope: how it becomes microscopic in times of trauma and loss; how it is no less present or powerful for being so tiny. Life and memory find a way to endure amid gross, frightening entropy, the near-total breakdown of the familiar. We can't make out the past or begin to imagine the future at this moment of dissolution, but thanks to Dr. Weiss and other biologists' work over the past sixty years, we know that the seeds of it are in there, floating in the muck.

We like to frame the world in stories, and we tend to think they should progress logically. Yet the butterfly shows us otherwise. It reveals that, much like our own realities, a story can dissolve into goo in the middle and end up as a completely different animal. And in that, it is also a microcosm for so much of science as a whole. Sometimes we get exactly what we expect. But much of the time, we don't. Science has a way of ensuring reality intrudes, whether we like it or not, and ultimately that's a good thing. It helps us evaluate our lives in the context of the world that is, not amid the false worlds we are prone to construct.

Among its lessons, science shows us that we live in a world that can't handle unchecked growth. Ever. Our bodies aren't meant to handle it; our planet isn't meant to handle it; and our civilization can't sustain it, whether we're talking about the economy, development, or anything else. We've spent two centuries—initially in the West, now all over the world—trying to pretend otherwise, and look where that has gotten us. Scientist and author Hope Jahren aptly labeled it *The Story of More*, but sooner or later the world requires stories of less. It was never true that growth can just keep rolling along. And that's why the caterpillar has something so important to teach us: on this planet, healthy growth, hopeful growth, involves disintegration as well.

My own moment of disintegration arrived on a gray November day, sitting with my wife, Diana, in the radiology waiting room of a Colorado children's hospital, her hand threatening to crush the bones in mine. Our daughter, Neva, still lay somewhere beyond the double doors and their Do Not Enter warning. Every atom in our bodies struggled to break that command when a pediatric oncologist with kind eyes sat beside us and uttered those seven words about Neva's MRI.

More words followed. They included *tumor in the pituitary gland* and *will require brain surgery to remove.*

Somehow, I processed this information and even found a way to ask pertinent questions. I did this because while I silently begged and hoped and prayed that the news would be different, the realization this was coming was already lodged in my gut. The data pointed the way. A set of X-rays showed that she had the bones of a toddler. She couldn't do many of the normal things on a playground that her peers could. She was getting multiple headaches each month. None of it was normal for a four-year-old. Though cautioned against it, I took that information and googled and read and pulled articles from the primary literature until I arrived at yet another word: *cranio-pharyngioma*. In some terrible way, I was prepared. There are times when I wish the scientist in me would go away.

The oncologist had arrived in the radiology waiting room with a team of residents in tow. I remember wondering how they all felt as they sought us out. Did it become routine, this doling out of eviscerating news? Or did they approach a parent with knots in their own stomachs, perhaps deliberately welcoming in just a bit of the impending pain in hopes that it could lessen the blow, if only a little. The residents all looked unsure of themselves, perhaps still figuring out the complex emotional calculus of it all, but as I looked at the attending oncologist, I remember thinking, *This man does not leave his humanity behind*.

He told us the MRI tech called him while the scan was in progress, letting him know there was clearly a lesion of some kind in my daughter's brain. That gave us the chance to absorb the initial blow without Neva present, but suddenly I had an overwhelming need to see her.

Can we go back now?

Yes, of course. And please remember that we still need the official radiology read. But I'm nearly certain this is a cranio. A follow-up CT scan will confirm. Go see her and get some food, and then come up to the seventh floor and we can discuss next steps.

A nurse ushered us through the doors and along a row of mobile beds. A boy of about twelve lay in one, with sullen eyes fixed on the TV. A baby was in the next one, wailing as a nurse bent over the bed. Neva was in the last bed, an IV still in her arm, electric leads from her chest to a monitor above. Her thick brown hair was a tangle, her blue eyes open but glassy.

Mama.

My wife brushed back a lock of her own dark hair and laid a hand on Neva's chest, kissed her head, and told her everything was okay. The similarities between the two of them were striking. Big eyes that danced with deliberate mischief, enormous smiles, minds that typically raced ahead of everyone else. A constant tendency to question the standard narrative. As Christmas approached the year before Neva's diagnosis, when she was only three, she sprang a trap on us in the form of a logical proof.

Mama, how does Santa get on the roof?

Well, he flies there with his reindeer.

But isn't flying kind of magic?

Well, yes, I suppose it is.

But, Mama, if Santa is magic and magic isn't real, then Santa can't be real.

Even at birth, Neva emerged from her mother with a calm and skeptical look beneath a shock of dark hair, as she regarded her suddenly expanded world with an air of critical assessment. The delivery nurse had to force her to cry.

There was a double frame on our mantel at home that held pictures of Diana and Neva at the age of three. Neva wore a bright red dress with stripes on the sleeves, while the other picture of Diana in a patterned jumper was faded by the years. But absent the markers of time, the two little girls were nearly indistinguishable. Friends would look at my wife and daughter and make jokes about an immaculate conception, or perhaps say something like:

Clearly biology is wrong, and genetics is not fifty-fifty.

Sometimes following that with a smiling, *Thank God.*

Diana turned to the nurse, who stood beside a softly beeping monitor on the other side of Neva's bed.

Can she eat and drink?

Let's start with some ice chips and make sure those go down okay. Then she can have some juice and crackers if she wants.

My daughter held the saltines in her hand, crumbs coating the little hospital gown with its repeating patterns of dogs and cats. She sat up and looked at the port in her left arm, encased in clear tape.

Dada, what is this?

Scared, her voice rising.

It's called an IV, honey. It helped make sure your body got what it needed while you were asleep for the pictures.

I want to take it off.

The nurse will very soon.

Will it hurt?

Just like a Band-Aid, honey. It won't be hard.

Band-Aids hurt to take off!

The nurse cut in.

I have a special cream I can put on that to make it really easy.

Can I see it?

The nurse handed Neva a white tube, and she studied it carefully, turning it back and forth. She was back to her critical evaluator self.

Okay.

Soon the IV was out and we helped her put her clothes back on, then I carried her through the brightly lit hallways of the recovery ward before emerging into the softer light of the waiting room. A small café with glass-walled ice cream bins beckoned from just beyond. She chose mint chocolate chip, stained her white shirt with a blot of green, and asked if we could go home.

Not quite yet. We need to go upstairs and talk to the doctors a little. And you may have to get another picture, but this one will be quick and easy.

Do I have to get the needle in my arm again?

She had a defiant look in her eyes.

No, love. And we can be with you. All you have to do is lie on a table for a couple of minutes.

I want to go home.

We do too, and we will soon.

Ever the scientist she was, Diana had her laptop out while Neva ate her ice cream, and was searching for papers on the treatment of childhood craniopharyngiomas. She scooted her chair around to show me the abstract of one of them, a study assessing whether or not radiation should follow surgery in younger children. The answer: at Neva's age, maybe not. My wife was in a mode I knew well—completely focused, undoubtedly stepping into research as a way to deflect her own stress. But I wasn't ready to join her yet. We whispered back and forth, trying not to let Neva hear.

We don't even know yet for sure that's what she has. It could still be something different, maybe even fully benign.

Me, grasping for a foothold of denial.

You heard the doc. She had a slightly accusing look. *I'm trying to get ready for the conversation we are about to have.*

I know, I know. But maybe let's just wait until we know for sure?

I could see Diana was skeptical, but she closed her laptop anyway with a brief sigh. She took my hand, held it momentarily, then put both of hers on the table. She drummed her fingers and shifted in her seat. Then she stood up and walked to the café counter, ordered a drink, paced back and forth while a kid with quarter-sized discs embedded in his earlobes fired up the espresso machine. I realized that steering her away from the treatment research was a mistake, so I walked over.

Hon, maybe it can't hurt to read a little more before the meeting upstairs? Be more prepared just in case?

She looked at me gratefully and returned to the computer. Neva asked if she could have more ice cream.

Hell yes, I thought.

She still carried the ice cream cup as we rode the elevator in silence to the seventh floor, then approached the oncology department's waiting room. Its bright colors could not lessen the impact of the girl in one corner, her head bald and her arm connected to an IV mounted above her wheelchair. She was a couple of years older than Neva and rail thin, with resigned eyes that her gaunt face made all the more prominent. We tried to distract Neva as she stared. Tried to distract ourselves. Finally, a call came from the desk.

Neva?

I was told that we could return to radiology right away for the CT scan. The MRI could display the tumor itself in stark detail but could not see the high concentrations of calcium that were diagnostic of a craniopharyngioma. The CT could. And it did. An element I measured in my own laboratory now exploded the last shreds of hope for a different outcome.

The oncologist delivered the news slowly and evenly, with a practiced calm and kindness. He said something about a surgeon named Todd who was "just amazing," about a prognosis that was still hopeful, about how this was just the first step on a long journey. All of it mostly bounced off my brain as I looked at the office walls. There were stickers of Disney characters on two of them and a bulletin board on a third, overflowing with cards and photos. Children with bald heads and large eyes but smiling faces. One was gamely holding a baseball bat as he stood at the plate in his Little League uniform. My eyes fell upon a note beneath the picture thanking the doctors for doing all they could. Even though it wasn't enough.

A few years before, Diana and I had sat on a remote Costa Rican beach, watching pelicans cheat the waves. The birds are surprisingly graceful in flight, their wing tips often brushing the backside of rising swells in angled lines, mini fighter jets come to life. Typically, they would wheel away from the breakers, perhaps rising up and then falling into a sudden dive as they took fish from the surface. But on this beach, huge schools of mullet were pinned to the shoreline, and the pelicans were rolling the dice. Diving and eating just inside glittering blue walls of five feet or more, then taking off at the last minute to barely escape. Until one pushed it too far.

The unlucky bird rose up the face of the cresting wave, furiously trying to become airborne, but its wings caught in the foaming top before it was slammed violently to the water below. Both Diana and I recoiled at the sight and looked on sadly as the seemingly dead mound of brown and white feathers washed toward the beach and lay motionless in the shallows.

Then my wife let out a small cheer as the pelican rose unsteadily from the whitewash. It staggered up the beach like a drunkard at closing time before coming to rest just a few feet away, staring vacantly back toward the sea. Diana turned to me.

Poor bastard. He has absolutely no damn idea who or where he is.

Walking out of the oncologist's office, I was that pelican. Nothing would come into focus. I felt as though I would, at any moment, simply shatter into a million brittle shards of something that once was but could no longer be. Like so many distressed caregivers before me, I began to drive myself to near insanity by asking the question:

How could this possibly happen?

How can my daughter now be the one harboring dangerous and unchecked growth? In her case, the craniopharyngioma didn't threaten to metastasize and unleash that growth all over her still-tiny body, but because the tumor was in her brain, local growth alone was a huge risk to so very much. Her eyesight, her basic ability to regulate a host of critical bodily functions, even her life itself. It was overwhelming, and I couldn't stop thinking about how this was not supposed to be her story. For years, I'd stood in the front of classrooms and told students about other forms of unchecked growth, about their risks, about how we needed to change the trajectory our planet was on. But in truth, it always felt a bit more intellectual than personal. Now those moments came flooding back in a very different light, as I thought about how desperately I wanted to stop what was growing inside Neva's head.

Sometimes I'd start by telling students about a common approach to unraveling stories in my field that is known as substituting space for time. In effect, it means that because we can't just wait around for years or even decades to see what a given forest or pasture or lake might do under a certain set of conditions, we have to rely instead on examples of that progression spread across multiple locations. That's exactly what Greg and I were doing in Brazil: finding pastures of different ages, and from that, reconstructing a tale about how soon after the forest was burned—and why—a given pasture would fall apart.

But those same Amazonian forests are also an example of a different space for time swap, one society as a whole is using to fuel our grand self-deception. In the blink of a geologic eye, we've gone from seeing our insatiable penchant for growth play out on largely local scales to one where we are filling that demand from every corner of the globe. Most of the new soybeans sprouting in places once occupied by those Brazilian forests were shipped across thousands of miles of land and sea to feed pigs and chickens in China and the United States. To keep growing in one place, we must destroy growth somewhere else, and we have accelerated that destruction right off the charts. But we've become enormously skilled at hiding that reality, thereby allowing us to keep pretending all that growth can simply keep going. That it should remain the organizing goal for each of our lives. This is our dominant story.

Now I understood our collective denial in a new light, because I wanted nothing more than to hide from a far more personal confrontation of growth's inevitable limits. I wanted to go back to the story of Neva from before. The one of a big-eyed beauty who liked to sprint the hallways of her preschool wearing a dozen costume animal tails at once: a zebra on the left, a lion on the right, and an Ark's worth of others crowded in between, singing that zydeco song "Audubon Zoo" at full volume like a tiny reincarnated jazz man from New Orleans. That Neva was supposed to grow up in the delightful and maddening and utterly ordinary tangle of girlhood before heading out to grow her own life, accumulating things as we all seem hardwired to do: degrees, jobs, partners, friends, families.

As we drove home from the hospital, I knew at least part of that story was now gone. And I hoped and prayed that, much like the butterfly, its dissolution would somehow transform into a miraculous new reality I could not yet see.

CHAPTER THREE

More than a decade before that painful day, I faced another wave of stress and uncertainty as a series of events began to unfold that eventually would lead me to Diana. I was scrambling for control over my professional future because Greg and I had been kicked out of Brazil. Almost overnight I was confronting the reality that I'd have to restart my career in a whole new country.

Kicked out is perhaps overly dramatic. But our project had become a casualty of growing paranoia in the Brazilian legislature. The work was funded by NASA, which had much grander plans for a scientific mission all across the Amazon. Then a few Brazilian lawmakers cried foul, accusing the enterprise of being a smokescreen for US extraction of Brazil's sovereign resources. Arguments and posturing ensued. Laws were passed. Among them: no scientific samples could leave the country. No soil, no plants, no water, not even a vial of air. Try it and risk prison, as happened to at least one PhD student who was desperate to finish his degree.

The Brazilian project was my first one as a new professor at the University of Colorado, and it formed the center of all my plans. Academic science is a funny business; you pour everything into landing a job, usually against long odds, so if it happens you feel a ton of relief. But that respite is brief, because you quickly pivot to what it will take to *keep* that job. You've got six years to assemble a successful research record. That might sound long, but those years fly by, and given the time it takes for every step in the process—setting up your lab, writing grants, recruiting students, collecting data, analyzing those data, writing them up in papers, waiting months or more to learn if those papers will be published, all while juggling a hundred other demands for your time—you'd better hit the ground running.

The Colorado job felt like winning the lottery, so the six-year deadline loomed even larger when Brazil became a dead end. Without the ability to bring samples home, I couldn't finish the current project as proposed, let alone build the longer-term research program I envisioned. The specter of tenure denial hovered. So did feelings of loss and sadness. We were starting to build relationships, to get answers, to learn about a place that was changing fast and needed those answers. I wanted to know the end of the story, but now the book was yanked from my hands.

Spend enough time doing science, and you learn this kind of stuff is the rule, not the exception. Failure is common. The entire enterprise is built upon peer critique, sometimes constructive, sometimes not. Most grants and papers are rejected. Things break in the lab and in the field. Samples get lost, ruined, or just don't work out as you expect. Somebody else publishes the idea you've been cultivating for the past year. The hits come often, and that's if you're treated fairly...and many are not. Because in the end science is human, meaning its inherent challenges are deepened by our own failings.

In other words, the story doesn't always progress in the ways you envision. Like the butterfly, science teaches you to expect repeated dissolutions of the narrative you think you're in—and hopefully the ability to adjust to the one that now is. Sometimes you never get out of the muck. But every once in a while, a remarkable new reality emerges that you could never predict.

After the Brazil news landed, I spent hours poring over maps of different parts of the world, trying to figure out where I could reboot the project. After considering a few options, I began to focus on the southwest corner of Costa Rica. On those maps, the place kind of resembled my misshapen second toe. The toe bends one way, then the other, as though seeking an escape—as though convinced it does not belong with the rest of its foot-mates. So too is this chunk of land disconnected from its neighbors. Known as the Osa Peninsula, the jungles draped across its corrugated landscape are bigger and wilder than those in the rest of the country and much more closely related to ones you can find in the western Amazon.

And, like the Amazon, some of that forest was being cut down. So, I pulled up stakes in Brazil and hoped the Osa would keep NASA happy . . . and me my job.

Greg moved on to other pursuits, so about a year after Brazil closed the door, a topo map of the Osa was in the hands of a new partner with a tangle of strawberry-blond hair and a distinctive and infectious laugh. Cory and I first met years before, when I was a graduate student, him an undergraduate, the two of us in a basement laboratory at Boulder's National Center for Atmospheric Research. Now we sat beside each other in the back of a dented red taxi that was rocketing through a suburb of San Jose, Costa Rica. Cory studied the map while I watched the approach of a narrow runway lined by a chicken-wire fence and meter-high elephant grass. Soon I could see a scattering of planes on the tarmac and a couple of hangars. Cory had become my first PhD student, and Pavas Airport was our portal to a small hotel on the northwest shore of the Osa. From there we'd range by boat and foot to explore the forests of Corcovado National Park and beyond. But first we found ourselves in one of the dilapidated hangars negotiating with a pilot.

Señores, is too much luggage. You cannot have more than twenty-five pounds.

Yes, I know, but I called in advance. We're biologists and need to take this equipment with us for our work.

He looked at us for several seconds without saying anything, then walked into the charter office and began gesturing with another man, both of them in matching dark pants, starched white shirts, and thin black ties. After a few minutes, the men emerged from the office, looked over our pile of duffels and action packers, muttered to each other in Spanish. Then they returned to the office and sat at separate desks, drinking coffee. Meanwhile, we leaned against a luggage cart in the hangar and waited them out. If nothing else, fieldwork teaches you to be patient.

When the second man finally emerged, he simply said:

Okay, we go now.

We squeezed our gear and ourselves into a sun-bleached Cessna with a single engine up front. As we flew west toward the Pacific, tracing the coastline to the south, the northern edge of the Osa Peninsula came into clear view. I could see a spine of forested ridgetop extending to the ocean, punctuated by two rugged islands just off the ridgeline's entry to the sea. Next, a shining bay with another much smaller river entering at the midpoint of its crescent shoreline. As we passed the river, the plane banked east, revealing a scattering of buildings and pastures just inland, along with the southern point of the bay, where we could see manicured lawns amid maybe two dozen thatched roofs. This was the rustic hotel that would become my second home for years to come.

The plane leaned into a final and much sharper turn before we touched down on a gravel strip. I remember the stress of the prior months melting away in those final moments before landing. My curiosity was taking hold again, and I couldn't help but wonder what we'd find in this place, how these forests might work, what new answers might come. I felt the same kind of relief I'd seen on Manuel's face as we stood in his small Brazilian creek.

By the next morning, the feeling only deepened as we explored the region for the first time. The pastures just beyond the airport, as well as the forests on the hills above, all belonged to the hotel owners, and their land was our primary target. Reaching it meant a jarring ride in an ancient forest-green Land Cruiser, where two absurdly tiny fans wired to the ceiling tried—and failed—to keep us cool. The man driving was tall and dark with an angular jaw beneath a fraying straw hat and a noticeable limp that he still managed to make look like a swagger. He had introduced himself as Hilberto, the name coming out in a drawl with a thin smile. A Costa Rican John Wayne.

The road came to an end in front of a slightly sagging one-story structure with three entry doors and a red metal roof. To that building's left, there was a nicer home of stucco and Spanish tile. Three skinny dogs greeted the Cruiser's arrival with lethargic barks while Hilberto climbed out and walked a few steps toward the decaying edifice. This was the workers' quarters, where the men and women who ran a farm that supplied the hotel with food lived. The stucco home belonged to the hotel owners, but they rarely used it, and before long we'd turn it into a mini field station for our work. Hilberto climbed out of the Cruiser, made a few introductions, then led us on a hike all about the farm and into the stunning forest beyond.

We came back only weeks later, this time swapping the little Cessna for a diesel HiLux pickup that could haul all the gear we needed to launch the new phase of our research in earnest. After a couple of wrong turns, we escaped the frenetic confines of San Jose and headed south on the Inter-American Highway. Cory dug through a small box of CDs and fed one into the truck's dated stereo system. A countrified version of "Gin and Juice" blasted from the speakers as I downshifted to pass a spewing truck that was crawling up the initial slope of the Talamanca Range, the vehicle's high and wooden-planked sides barely containing a tangle of old furniture and twisted metal. It was the first of many trucks we'd pass, and I began to realize why a drive of only two hundred miles would take us the entire day. I settled in and told myself to simply enjoy a slow-paced tour through a new country.

Costa Rica is an example of what can happen when a continental plate goes crashing into its neighbor. One plate is shoved beneath the other, as if in some infinitely slow but titanic battle. The loser takes its revenge by melting as it is pushed ever farther toward Earth's core, and the resultant magma bubbles skyward through the overlying plate to form volcanoes. In geologic terms, the process is known as subduction. The deeper plate also continues to push back even as it is forced below, uplifting the land above. Costa Rica has some of the highest rates around, with some areas moving skyward about the length of your fingernail each year. In the uplift world, that's hauling ass.

The youngest volcanoes are found up north, some of them still active and easily recognizable by their conical shapes. Things are less obvious in the high backbone of the Talamanca Range, which bisects much of Costa Rica, because its mountains are the eroded roots of volcanoes that once sat upon the ocean floor before erupting themselves into dry land. That process ebbed and flowed over millennia, including times when land that had escaped the ocean found itself submerged once again. The eruption days are long over, but the range is still the high point of the country, parts of it above ten thousand feet and one high peak cresting to twelve thousand. There are scars of glaciers up here, where the trees gave way to a Seussian land known as páramo, which occasionally sees snow. You'd never guess that the rocks just beneath are replete with evidence of an ocean long past.

Like the Talamancas, the Osa began its life as a seafloor volcano more than seventy million years ago. As climates changed and the sea level rose and fell, it slowly worked its way toward the surface, along with the entire arc of Central America. The bulk of the peninsula itself, including all of Corcovado National Park, was once again consumed by the ocean more recently, but the northern arc stayed dry the entire time. Look closely, and you can tell the difference as you travel through each part of the region. Up north, the ridges are steeper, the road cuts redder—these are the marks of millions of years of erosion. Farther south, the landscape transitions are usually a bit gentler, while the road cuts are pockmarked by fossilized remnants of marine creatures caught up in the sediments that blanketed the southern peninsula during its final time beneath the sea.

Those differences were part of what brought us here in the first place. In theory anyway, the soils up north had been weathering away far longer than those to the south, each year losing key plant nutrients to the sea; that in turn should make them less fertile than those in the south, though likely still better than our former Brazilian sites. The idea was a multisite, multicountry comparison: What would happen after deforestation in the north and south here? How would the regions compare to each other, and to our sites in the Amazon? And could all of that tell us anything about how to slow the cycle of forest destruction?

Many tropical forests have evolved to be remarkably good at mining their ancient soils for what they need, then holding on to scarce nutrients for dear life, generation after generation. They pull those bits of stardust out of dying leaves before they fall to the ground and use a tangle of fine roots and fungal partners to quickly regrab portions that do escape the canopy. But cut those trees down, burn them, then ship regular truckloads of those nutrients away in the form of cows or soybeans, and problems arise in a hurry. It becomes a bank account that is quickly draining. Hence the rapid decline of the pastures we studied in Brazil. Here in this southwestern corner of Costa Rica, where even the oldest-looking soils seemed to have more life in them and where pasture burning was far less common, maybe the story would be different.

We didn't know yet. But as with Brazil, we hoped to connect what we measured on the ground to what could be mapped from the air—something we would do years later thanks to that custom airplane of Greg's design. Just like the Amazon, we had to begin by getting samples of soils and vegetation, then take them home for chemical analyses. But here we also wanted to measure and map the forests themselves. And that could sometimes become a theater of the absurd.

Take, for example, a day when we wandered through the forest repeatedly shooting an ancient shotgun at the sky. Why? Because we needed leaves from high up in the trees, and short of training a set of monkeys—something we half seriously discussed at one point—the shotgun was our best bet. I'd first tried the technique when I was a graduate student in Hawaii, where the trees were a lot shorter. There, you could line up a group of leaves, pull the trigger and most of the time watch the leaves you wanted float to the ground. On the Osa, things didn't go as smoothly.

The first day we tried it, I stood beneath a tree that was well over a hundred feet tall and tried to pick out a small branch that the shot could break free. I mounted the gun awkwardly to the top of my shoulder, then pulled the trigger, wincing as always at the impact on my clavicle. Cory and I looked up, watched the branch with its assembly of leaves release from the tree and begin to fall in oscillating arcs, as though it were a kite that was losing its wind.

Goddamn, it worked! I exclaimed.

Then the leaves landed in the top of another tree, still far out of our reach. Cory swore.

Shit. Try again, I guess.

I did. The same thing happened. And thus began a day where we ran through box after box of shotgun shells, battering our shoulders as we walked through the forest muttering curses and speculating about the monkey option. By the time midafternoon rolled around, I'd lost count of how many shots we had fired to assemble a depressingly small number of zip-lock bags containing the felled leaves. Then things came to a head when a pair of blue-uniformed officers accosted us near the edge of the forest. They were from the Costa Rican equivalent of the FBI, and a neighboring landowner had called them, convinced that a drug war had broken out next door. The two men's faces were hard and skeptical as we explained what we were doing, until we fished out a few of the bags of leaves as proof. Then they began to laugh, and one said something in Spanish to the other that we could not make out. But the meaning was clear. Idiot scientists.

We needed the leaves for the same reasons Greg and I sampled those pasture grasses in the Amazon. They contained elemental clues about how the forest functioned, clues that perhaps we could one day measure from space. In time, we dialed in the shotgun method just a bit better, made a shoulder pad out of an old piece of mattress foam wrapped in duct tape, and began to put together a new story. We learned, for example, how the forests would drop some of those leaves in the dry season, and then, when the first spring rains came, the water would release the nutrients from fallen leaves into the soil in a naturally brewed tea that seemed to light the whole place up. Bright green new leaves would appear in the canopy, followed by splashes of multicolored flowers. The animals would begin to move in ways they did not when the sun beat down day after day. All of it because those starborn elements were just doing their thing—moving from one place to another, fueling another season of life.

And yet, while my professional life found purchase, my personal one was falling apart. Sometimes I think that a basic rule of science often applies to life itself: a version of what's known in physics and chemistry as the conservation of mass. That rule states that in a closed system, you can move energy and matter all over the place, but you can't change the overall mass. Put some more energy into one spot, you have to take it away from somewhere else. So it seemed to be for me. A marriage that had warning signs from the beginning was now unraveling for good. We'd separated twice before, tried to come back together, and had two children, all of it amid an odd pairing in which we spent six of our eight years living in different states.

The weight of it all dragged me down into what was, at the time, the low point of my life. But like that butterfly, science teaches you that a moment of contraction and chaos—a time when everything seems to be falling apart—is probably temporary. That something surprising and wonderful might lie ahead. You just have to keep yourself in the position to find it, and as Diana would ultimately show me more powerfully than anyone else, perhaps discover some wonder and peace in not knowing how the story will end.

I don't think that's how much of the world looks at science. There is a moment in John McPhee's *The Control of Nature* where the author reflects on a battle against an Icelandic volcano, writing: *In making war with nature, there was risk of loss in winning.* The line has stuck with me over the years because, as was McPhee's intent in the entire book, it suggests something deeper about how many people view science and, by extension, ourselves. That it's a way to find answers that will help us exert greater control over the arcs of our lives.

It's an understandable allure, because science *has* reshaped our world in ways that were simply unimaginable just a few generations ago. And that creates daily illusions of control, every time we change the thermostat, take a prescription medicine, drive over a bridge, or even call 911. It all reinforces the belief that science is a path to answers that, in turn, erect walls against all manner of uncertainty and danger. Which, like any good self-deception, is a belief that is anchored in elements of truth. But to me this view of science is not only somewhat misleading, but it misses some of its greatest power.

Perhaps more out of habit than from any conscious decision, I started to let the practice of science during those early years in Costa Rica carry me through a shitty time. I remember standing in the forest one day reflecting on how much I still didn't understand about how that ecosystem worked—and how fun the uncertainty was. Then I thought: maybe I could try to look at this moment in my own life the same way. I couldn't control significant parts of what was unraveling in my life. And I couldn't know much of what was to come. But I could try to find some pockets of joy in the not knowing, and I could remind myself that sometimes the eventual answers are better than we think.

In that, what was first an escape from the stresses of my home life began to morph into a place where I could face some painful decisions with a little more peace and clarity. On one of those trips, Cory and I stayed up late into the night, first just talking about life and then hatching the next phase of our research plan. And it worked beyond my wildest dreams, because it led me to the woman I'd come to love.

CHAPTER FOUR

I met her because of another setback. That new research plan Cory and I designed required that we bring in an additional collaborator—a more senior professor in my department. We spent months planning a critical field campaign, and then just a few weeks before it was time to go, he bailed. The news was delivered as Cory and I sat opposite his desk.

Yeah, sorry, guys, just can't do it. But I can send one of my PhD students instead.

Cory and I shifted in our chairs and stole a quick glance at each other. All we knew of this student was that she appeared to be a lab rat with no field experience. We'd soon learn that she had never even left the country. I tried to find a way to delicately push back.

Um, you sure? This is pretty wild and remote country, lotta snakes, kinda crazy sometimes. And we have a hell of a lot to get done on this trip. Maybe not the best time or place for someone's first real field campaign?

Later I'd come to understand the slight, knowing smile that appeared on my colleague's face.

Trust me, Diana's tough. No doubt she'll be way better than me.

I was too distracted to argue more. Soon enough, the window of the airport bus showed a blur of suburban sprawl, and I was only vaguely aware of the one-sided conversation to my left. My newest student, Sasha, a hurricane of energy and enthusiasm, careened from one topic to another while Cory and Diana mostly listened and laughed. A few hours later, after our flight to San Jose was canceled, the four of us sat on rotating stools at the bar of a nondescript Dallas airport hotel, while a reception for an Indian wedding filled the room and hallways. A color wheel of saris and lenghas. I watched three rebellious teenage boys out the window behind the bar leaning against a stuccoed wall, one holding a cigarette low and tight beside his right leg. Endless loops of bhangra pushed through the thin wall to our left. The music and the conversation of my traveling companions wove together into white noise as I remained locked in an internal cycle of depression over my imminent divorce. For some reason I ordered a drink I rarely chose: a martini.

Martini? Forget your ascot?

Her smile was enormous, at once welcoming and gently challenging. Her eyes were heterochromic and lit. I tried to shake free of the hole I was in and respond.

Well, they're all out of port.

You don't have your elbow patches anyway.

Clearly, I was going to lose this battle of wits, and I didn't have the energy for it anyway. I told her as much.

I'm sorry, I'm not much fun right now. Kind of a tough time for me.

Diana looked at me for several seconds, as though trying to decide whether my Eeyore line was an invitation or a dismissal. Then came a look of startling kindness.

It's okay.

The words came slowly and with an empathy I could feel. She looked at me for a few beats more, then turned back to Cory and Sasha. I sat and looked at my drink, smiled, and for a moment felt buoyed. But it didn't last. I slipped away and spent an unknown block of time sitting on a cheap Naugahyde ottoman in a dark hotel room, staring dully out the window. When Cory arrived and flipped on the lights, I could see the concern on his face.

You okay?

Not really. But I will be.

My fugue continued through our arrival back to the Osa. I'd come to love this place over the years we'd worked here, but as we prepared for our first day in the field, I still couldn't shake my sadness. I tried to hide it, but Cory knew me well enough to see through the facade. He raised an eyebrow when I suggested he, Sasha, and Diana go one way into the forest, while I went to check out another area.

Okay. We'll hook up with you in a bit.

I watched them walk away and then turned toward the other direction, and as ever, it was like entering a cathedral.

The trees in these forests are astonishingly varied—at least seven hundred different species live there. One vaguely resembles an oak yet drops a carpet of yellow flowers each year that makes the entire area smell like garlic. A few others swell grotesquely at their base, as though stricken with some form of arboreal gout. Still more are covered in reptilian armor, earning them the name "alligator tree," while another, the cow tree, exudes a milky sap from its dappled trunk that locals swear will cure all manner of ills. Among my favorites is the zapatero—Spanish for shoemaker—whose giant yellow roots wind along the forest floor and whose trunk will break just about any chain saw.

One thing unites all of these trees and all the others on our planet: they live life where it is and as it comes. If you're a botanist reading this, you'll know that's not entirely true—technically, a few trees actually *do* move during their lives. In the Osa forests, you can find walking palms, spiny little horrors that creep their way around the forest floor by sending out new aboveground roots in the direction of the most fertile patches of soil. These fuckers once left part of a two-inch thorn embedded in my skull for weeks.

Like me at the outset of the trip, walking palms are defensive and ugly, not content with where they stand. The parallel struck me as I walked farther into the forest, at which point I sat on my backpack and held back tears. Was I a shitty, spiny little escape artist, leaving my marriage while I sought something better? I didn't know.

By now I'd spent countless hours observing these forests, seeking answers to how they worked. So after a few moments of beating myself up, I began to focus almost by instinct on the sweeping branches of a tree known as a *chiricano*, recognizable by its coarse layering of purple bark that seems almost like an oil painting come to life. On a prior trip, this same tree held me in its central fork as warring troops of monkeys surrounded my perch while I was trying to saw loose a few leaves. That had been during a brief interlude where we gave up on the shotgun and decided to try to climb the trees instead.

At one point in Joy Harjo's "Eagle Poem," she writes: *Breathe in, knowing we are made of all this.* On that day in the forest, I began to think of how the cycles of our lives, our emotions, our moments of struggle or acceptance kind of resembled the cycles of the elements I was there to study. Sometimes those elements are trapped in chemical prisons, such as the structure of rock or clay, from which there seems to be no escape. Sometimes they are ripped violently from their longtime homes, then deposited somewhere entirely foreign and full of chaos, like when a landslide suddenly dumps part of a hillside into a flooding river. And sometimes the elements are released into the air, pulled off the ocean in tiny droplets, or exhaled in the metabolism of life, where they float peacefully above the forest before raining back down to become part of the trees bathed in that rain.

Not that it was all sudden enlightenment at the time. A big ant with a yellow butt bit me on my peach-colored one as I sat in the dirt. I shot up with a string of curses and stomped the ant into oblivion. Then I felt bad for doing so. I walked back down the gentle slope toward where I thought the others would be, rounded a corner, and saw a group of capuchin monkeys clustered low in a nearby tree. They glared and chattered while shaking branches, and one of them threw shit at me, which at the moment felt just.

As I walked on, the sky darkened and a nearby howler monkey began a throaty roar. Others joined him, and I knew that was a sign of the rain to come. It began with staccato beats on the leaves far above, then soon enough became astonishing sheets. The downpour was so deafening that when the rest of the group found me just standing there getting soaked, we gave up trying to talk. We simply let the rain take over. It was baptismal: a moment where acceptance of the world as it was carried me to something extraordinary. I felt the forest and the mindset it triggered begin to dissolve my ego, as I simply observed, without judgment, without preconceived outcomes. I just stood, and I looked, and like Saint Paul's poor reflection in the mirror, my own self-image started to form into something clearer and more palatable when I simply let the love for what surrounded me take over.

In those moments, I began to see how a scientific mindset, at its heart, tries to teach you to put your ego aside. For most of us, that's no easy task. But if we hope to connect to something larger than ourselves—the very essence of spirituality—we have to get out of the traps our own minds want to set. On that day, as I stood in the drenching rain, I continued to let that forest show me how the lessons of my profession might let me bridge to a world of the unknown and find comfort there precisely because I don't have all the answers.

As the downpour continued, the four of us ran from the jungle back toward the truck, cracking up the entire time. We exited the forest near the workers' quarters, where Cory yelled out:

Hope that fucking Cujo is gone!

I laughed, finally free of my self-imposed prison, and answered him.

Yeah, and the damn chicken too!

Sasha broke in.

What's Cujo? What's the chicken?

We stopped running as Cory responded.

Cujo is a nasty German shepherd they had the last couple of times we were down. He's unpredictable. Walk by him three times and he doesn't care, the fourth he's looking to tear your throat out.

Oh great. And the chicken?

This time I weighed in.

It's a curassow. You know those big black birds, little yellow thing under their chins? Kinda like a turkey?

Ah yeah! I want to see one of those!

Well, you don't want to see this one. He's got a screw loose and for some reason hangs around the farm. When he sees us, he comes running and tries to attack if your back is turned.

Diana expressed doubt.

Bullshit. Come on. Some kinda wild chicken that hunts you down?

Cory shrugged.

Just watch your back, I'm tellin' you.

Then we started to jog again as the rain strengthened, failing to notice Cujo eyeing us from beside the building. As we passed, he sprang to life with a roar.

Screams, yells, sprinting. The curassow joined the fray seemingly from nowhere, dog and bird converging upon us from opposite sides. I swung a pack at Cujo while the bird landed on Diana's back with a squawk before Cory punched it aside, shoved Diana into the truck, and dove in behind. Sasha and I ran for the opposite side, jumped in, and slammed the doors. There was a flash of black feathers beside my window and then the sound of the curassow's feet scrabbling for a foothold on the pickup's roof before we sped away in a spray of dust and gravel, leaving the bird indignant beside the road. Inside the truck, we fell apart in laughter.

The next afternoon, the four of us sat outside our cabins following a sweaty morning of digging soil pits in the forest. We were still laughing about the dog-bird attack and enjoying a break in the rain. Eventually Cory and Sasha wandered off and then Diana asked, *Anywhere I can go running?*

Sure. If you climb the hill trail behind the office and go down the other side of it, then follow the coast for several miles.

Want to come along?

The rain hit just as we made the coastal turn, washing away our sweat and turning the run from hot to perfect. I followed behind her, trying to keep up. We ran for miles to where the trail ended at a beach backed by a rain-soaked lava cliff, where a patch of forest clung to the plunging slope, all of it framing the brilliant yellow of a single *mayo* tree in flower. There was no hint of humanity in sight.

The run back was a smiling but unspoken contest, each of us pushing the pace until we were hurdling over tangles of rain-slicked roots with abandon. Then the inevitable happened. Diana slipped on one such root, hit the ground hard, and rolled over while grabbing her left ankle. Her face contorted. I sat beside her as she held her knee to her chest, dangling her left foot in the air. Each time she set it back down, she grimaced anew. I spoke.

We're about three miles away. Want to lean on me and see if you can walk supported at all? If not, I'll run back and get a boat.

No, just give me a few minutes. I'll be okay.

I don't know. Looks bad. Really, I don't mind.

No, I'm fine.

She began to hobble down the trail, jaw noticeably set. I protested; she shot it down. Five minutes later she was running again. That night, the ankle ballooned noticeably beneath bag after bag of rapidly melting ice.

The swelling was still evident when she swapped field pants for running shorts at the end of the following day's work. I found her on her cabin porch, ripping an old T-shirt into strips.

She wrapped the ankle with them and overwrapped that with duct tape. Put on her shoes.

What are you doing?

Going running.

She had a look that challenged me to say more.

We ran the coastal trail again that day, then three more times as the first week bled into the second. Her ankle was duct-taped each time, and never right. It didn't seem to matter. With each run, bits of our lives leaked out, slowly at first, and then all in a rush on the final run of the trip. This one again in a dousing rain. Her own divorce. Her time at nineteen as a volunteer in the New Orleans projects, a world apart from her upstate New York childhood. Her brush with medical school, avoided only because the Tulane dean asked her if she intended to have children while there.

None of your fucking business, she'd replied.

As we walked slowly over the final hill from the coast to the hotel, I let the oscillations of my own history flow out. She just listened and held my gaze when I was done.

At the end of the trip, we had to split up the group. Cory and Sasha took an early-morning flight to San Jose, while Diana and I were left behind with the job of packing up our gear and samples and then driving it all to the city. Getting everything ready to go took us until midafternoon, so our plan was to drive partway back rather than battle the highway across the Talamancas in the middle of the night. We stopped in the coastal town of Dominical, where we booked a two-room cabin for the night.

We ate dinner at a small restaurant on the edge of town that had six circular metal tables, each with a white vinyl cloth. Our table contained the ubiquitous bottle of Lizano and a rack of worthless napkins as centerpieces. There were a hundred things I wanted to say and not one that would emerge. A boyish waiter in dark jeans and a pressed white shirt saved me.

Hola, buenas noches. Algo para tomar?

I ordered a Pilsen and so did she, and we drank the beers too quickly, then ordered more. Tinny pop music blared from a dated boom box on the high counter separating the kitchen from the dining area. By the time our casados arrived—grilled fish, beans and rice, salad of shredded lettuce and tomatoes—the conversation was flowing once again. She began to share more details of her childhood, looking uncharacteristically vulnerable when she described a German shepherd attack at the age of six. Then she laughed as she described the antics of a family named Strife, whose endless supply of giant boys tortured her hometown but protected her on the bus. From there, her jaw set just a little as she talked about a girl named Marcy who beat her out in the finals of the Miss Irish pageant.

I told her she didn't seem like she'd be into pageants.

I wasn't. But I entered because Marcy pissed me off.

After dinner, we took the two last stools at an open-air bar. The whole place felt just a bit on edge, but she gave me her shit-eating grin.

You gonna order one of those martinis?

I smiled.

Fuck off.

We ordered more beer instead and then halfway into them she flagged down the tica behind the bar. The girl gave her a thin smile and leaned on the counter, tendrils of a vine tattoo wrapping over her bare left shoulder and encircling her arm to the wrist.

Yes?

Do you have any cigarettes?

When they arrived, Diana pulled one out and gave me a challenging look, as if daring me to say something. I thought *hell with it* and asked, *Can I have one?*

You don't smoke.

No, but I don't think you really do, either.

Only when I drink too much. Bad habit from my New Orleans days.

Then she continued to tell me more about that time in her life. How she skipped her senior year of high school to do a bridge program at Clarkson University, where her talent for math had her bound for a career in engineering. Both the vulnerability and the set jaw returned when she described how it all went to hell in her second year after she was assaulted by a classmate, so she decided to follow a friend south, ending up in the inaugural year of the AmeriCorps program. Her eyes looked distant and sad when she revealed that in her first week in New Orleans, the pastor of the church that served as a host for her group was brutally murdered. But when she talked about the year that followed and the work she and others had done in the city's lower Ninth Ward, I could see the pride on her face.

We left the bar and walked beneath a line of trees to the beach, where the moon reflected off the breaking waves. Diana began to run toward the water, me close behind, until we were beside each other, lifting legs ever higher against the oncoming surge. As a larger line of foamy white approached, I dove in, only to surface on the other side and find her also floating only feet away. We dove again and again before walking slowly back up the beach, clothes dripping, winding our way back to the truck. Back at our cinder-block cabin, we lingered in the common room before saying a laden good night. I rinsed the salt off and lay awake for hours.

We made it to San Jose by noon the next day. But not before Diana told me to pull over near a roadside café high in the Talamancas and then came across the cab. Put her lips on mine. Fingers. In my hair.

CHAPTER FIVE

Scents are a powerful thing. As Robin Wall Kimmerer describes so beautifully in *Braiding Sweetgrass*, those of comfort and abundance, like the smell of freshly tilled earth, literally settle us down. That happens because the smells release oxytocin, a powerful hormone that bathes us in peace and bliss. Standing beneath a tree on the Osa known as a cedrón, I was enveloped in the smell of its dropped fruits. Though thousands of miles away, suddenly I was back in my Colorado home, grilling fresh peaches from the Western Slope. Kimmerer writes of how our noses can open up our minds to see what is truly around us, and from there perhaps respond to the world's inevitable assaults with calm and grace.

But smells can trigger the reverse too. On the day we went back to the children's hospital to meet with Neva's potential surgeon, my brain was overloaded by smells that evoked anything but comfort: screaming antiseptics, sweating parental panic, a distant little fart of cafeteria beef. My reactions to these vile stimulants weren't just overwhelming; they were physically corrosive. As a biologist, I knew that a fight-or-flight response this strong would unleash a cascade of physiological changes that whittled away at my evolutionary architecture.

That's because while our brains are undoubtedly a miracle, in some ways they're also kind of pathetic. Even after something like 850 million years of animal evolution, they are as much chaotic factory floor as they are the *Starship Enterprise*. Things are overcrowded and the machinery does not always work in well-aligned precision. There are different teams doing different jobs, all crammed into the same big room, and when there's an emergency, they all thrash around and run smack into one another.

Take, for instance, the department of olfactory processing—scent—and the department of emotion. They're situated very close to one another in the hippocampus. If you're a person whose life is ordinarily safe and happy, when something scary happens, you might notice that your sense of smell gets dramatically better. That's because if a fire alarm sounds in the emotion department and supercharges activity there, the same thing will happen in the smell department.

On the other hand, if you're stressed out all the time—if the alarm is going off in your inner factory 24/7, so often that the workers have learned to tune it out—your sense of smell is likely terrible. People who suffer from chronic PTSD, such as child abuse survivors, generally can't pick out subtleties of smell the way people without sustained trauma can. So, as we rode the elevator toward the neuro-oncology department once more, I knew the overcrowded factory in my mind might be able to put out a fire but that it wasn't set up all that well to heal the burn marks left behind.

In time I'd come to realize that habits my profession at least tried to cultivate—curious inquiry, observation, and thought before conclusions are drawn, the recognition that what you see in the moment might not be the complete story—all had neurological powers. That cultivating these habits might actually make my brain less reactive to a stressful situation and more able to stay settled as I thought it through. There's even evidence that in these hospital settings, putting up artwork or other visual stimuli that can promote curiosity results in better outcomes for everyone involved: patients, families, and the hospital staff. But at the time, I didn't know this. Nor am I sure the knowledge would have mattered in the moment. This was my kid after all, and so as with any fight-or-flight response, I wanted a fix *now*.

Should you ever hear the crushing words *your child has a brain tumor*, among the first of your churning gut responses will be: *get it the hell out of there*. And for craniopharyngiomas, that was the prevailing approach, then sometimes the surgery was followed by a course of radiation just for good measure. The tumors did not respond well to any kind of chemotherapy. But while radiation could slow or stop their march, radiating a child's brain is no small thing, meaning no easy decisions.

Cranios grow near the pituitary gland, which rests upon a bony structure just above the roof of your mouth known as the *sella turcica*. Literally, the Turkish saddle. Above that is a cavity of unfilled space. It's as though the pituitary got some of the best real estate in the brain, able to kick back in comfort and watch the show above while it goes about its own business... unless a tumor like a cranio comes along. If it does, it takes over that cavity and ruins the view, pressing into the pituitary first before filling up the rest of the open space and taking aim at the structures above.

The good news is that a neurosurgeon can get to that cavity, either through the side of the head or more recently through the back of the nose. Once in, the surgeon can oh so delicately scrape and suction away the offending mass until hopefully the entire thing is relegated to life in a trash can or pathology lab.

The problem is, cranios are shitty neighbors. They not only shove aside but also grab hold of the pituitary gland, the optic nerve, the hypothalamus, to the point where taking it all out can mean permanent surgical damage to the brain itself. And that's where technical ability must give way to a tough call, often in the heat of the moment, while overwrought parents wait outside the operating room desperately hoping the surgeon will return to say:

I got it all and everything is fine.

Any good surgeon will know that pushing too hard for complete removal can unleash myriad new problems. But leaving part of the tumor behind has major risks of its own. It's far more likely to grow again, far more likely that the child will now need radiation to stop it, far more likely that damage avoided for now will only get worse at some point down the road. By the time we were back at the hospital only a few days after Neva's MRI, we already knew our choices would have to bridge calculations of numerical odds and a realm of faith.

Nearly all doctors introduce themselves by leading with the title. *Good morning, I'm Dr. Smith.* That's okay, they've earned it, and many know their patients will find comfort in the authority and knowledge the title implies. But when this bald man of medium height, cowboy boots beneath his white coat and a gentle smile on his face began with an easy *Hi, I'm Todd,* before quickly turning his attention to the scared little girl on Diana's lap, I was already 90 percent sold. Soon, Neva steadied and gave him a smile of her own and then headed off to the playroom with a nurse, this time comfortable leaving both of us behind. And when Todd pointed at the margins of the tumor, explained the full suite of treatment options with remarkable clarity and honesty, talked about the need for good judgment before and during the operation, and then urged us to call other surgeons to hear their views, I was all in. So was Diana.

He wanted to access the tumor through the back of her nose.

I'll need to talk with Dr. Chan, the ENT surgeon who works with me on this kind of thing, but I think we can do it. She's small, but I don't think too small.

Diana asked:

How does it work, specifically?

Dr. Chan would make a small hole through the bone layer behind her nasal cavity. Then I take over. It's a short path to the tumor from there. When I'm done, he closes the hole by making a graft with fat from her belly. She'll have packing in her nose for probably five or six days, and she won't like that much. But it comes out easily when it's time. After that, we make sure there are no leaks and send her home.

I cut in.

Leaks?

CSF leaks. Cerebrospinal fluid. Happens in about a third of the cases, just because there is positive pressure against the graft from the brain cavity. It's not a huge deal if it happens, just means a longer stay in the hospital to let it resolve.

Then the big question.

Do you think you can get it all?

Todd paused, reading a new wave of stress in our eyes.

Look, you guys, this is just the start of a long road. Let's take it step by step. Neva has a lot going for her on this, so you have reasons to stay optimistic. If I do have to leave some behind, it doesn't mean she won't be okay.

Her surgery was set for two weeks out. A full analysis of Neva's pituitary function was now required, and that meant another return to the hospital, once again on a morning when she could neither eat nor drink. This time instead of breathing herself to sleep beneath an anesthesia mask, she had to pee in a cup—a team effort with a four-year-old—then sit on Diana's lap while nurse after nurse tried to draw blood and insert an IV. Her cherubic arms hid her veins while she wailed with increasing fear. After multiple failures, someone from the hospital's Life Flight team—a guy who routinely had to put IVs into critically injured patients aboard a helicopter—was called in. When he too was unable to make it work amid Neva's racking sobs, we nearly called it all off. Finally, a nurse from ICU was summoned and on the eleventh needle stick of the morning, the blood flowed and the IV line held. But not before she'd moved beyond crying to withdrawn resignation, her eyes dull and downcast, her shoulders slumped. It was searing.

Back home she slept, woke in fear of more needle sticks, slept some more. The signs of her ordeal were still evident that night as she oscillated between withdrawal and crying through pleas that she would not have to go back to the hospital. But kids seem to have a stunning ability to switch off the fight-or-flight response more quickly than many adults. It's partly a result of the plasticity of their still-growing brains, and for Neva, the switch seemed to have been fully flipped by morning. She woke up and hugged her dog, laughing as ever when her giant tongue soaked my daughter's face. She demanded a return to preschool to spend the day with friends and seemed blissfully unaware of how we lingered longer at the dropoff and hugged her more tightly.

A veneer of normalcy defined the next few days, and then it was time. A burnt-orange duffel accustomed to transporting field gear became loaded with picture books, coloring supplies, stuffed animals. A soft yellow blanket with repeating sea turtles in stitches of green. A scrimshawed music box with a tiny silver clasp. And a bleached segment of deer antler she had painted with glitter glue only months before, amid the sage and prickly pear of a high western desert. We spent the night at a hotel near the hospital, Diana and I sleeping little as we clung to her tiny form between us, before showing up to make a 5:00 a.m. check-in for surgery.

Somehow Neva strode through the sliding glass doors with her shoulders back and her head up, marching straight to the row of reception desks, with us trailing behind. We shook our heads in astonishment.

Mama, I think we go here.

Yes, we do.

Forms, signatures, a bracelet for each of us. Then it was up the glass elevator—*Mama, can I push the button?*—and into the surgery waiting area. Another check-in, this one quick, before we all sat on a couch of fuchsia Naugahyde. Waiting. Stressing.

He walked into the waiting room dressed as he nearly always was. Gray running shoes below quick-dry field pants of a nearly equal shade. The oversize dive watch and the black T-shirt. Three cups of coffee trapped between his large hands and an atypical seriousness in his eyes that was not masked by his usual grin. Greg. He handed us each a coffee.

I figured you two losers couldn't really handle this and that I'd better be here for my goddaughter.

Diana didn't miss a beat, but there were tears in her eyes as she shot back:

Your wife kick you out again?

I couldn't say a word. I just stood there as he grabbed my shoulder and said, *I know, man. I know.*

Greg often approached science with an almost ruthless drive and efficiency that evoked his Special Forces background. He had landed a position at Stanford University, and at one point as we compared notes on our approaches, he became impassioned about how important it was to construct clear boundaries between himself and the people on his research teams. To essentially hide good portions of his humanity and reserve their full picture for only a select few in his life. A military approach. And yet the irony was, Greg's heart was as big as anyone I knew. When he decided to let that lead the way, he was every bit the singular force he could be in science as well.

Neva Townsend?

A woman in blue scrubs scanned the waiting room.

Greg scooped our daughter up and spun her around before saying, *I'll see you after you wake up. Make sure your parents behave themselves.*

Moments later she sat on a gurney once again, an oven-warmed cotton blanket across her lap, the gown with cartoon puppies falling from her shoulders. She was unruffled by the squeeze of the blood pressure cuff, smiled slightly when the adhesive oxygen sensor turned her index fingernail a glowing red, and even laughed out loud when we donned Tyvek suits, shoe covers, and hairnets all of a matching shade of blue.

Dada, why are you wearing those?

So we can come into the surgery room with you, honey.

Do I have to have more shots?

She clutched the blanket tighter.

No, you'll be asleep.

Do I get to color on the mask again?

On the day of her initial diagnosis, in one of those small moments that can mean everything, a child psychologist who seemed barely out of college sat beside a similar gurney, patiently unveiling piece after piece of medical apparatus, giving Neva as much control over each one as possible. Near the end, she handed over a cup full of lip balm flavors, explaining that Neva could pick any one she liked and then color the inside of her anesthesia mask.

You get to pick the yummy flavor for falling asleep!

The psychologist was drawing on the power of scents. By giving Neva the chance to choose one that was both familiar and welcome, she knew it would elicit a calming hormone bath. But she may not have known an irony specific to Neva's tumor: if it grew further, it could shut off the very part of the brain that produced the oxytocin that made the whole thing work.

This time Neva chose strawberry. Todd walked in as she finished coating the inside of the mask.

Hi, kiddo. Did you make sure to get your mom and dad some coffee this morning?

She nodded silently, but hints of a smile pulled at the corners of her mouth. The nurse handed us more forms to sign, while Todd leaned close to us and whispered, *She'll be fine.*

Then he turned to Neva with a bright, *See you in there, kiddo.*

Twenty impossibly long minutes later, the nurse kicked free the foot brake on the gurney and began to push it down the hallway. Diana and I walked on either side, each seeking to have a hand on Neva at all times, trying to make small talk with her.

Honey, look at those dinosaurs on the wall! Hey, is it fun to get to ride on a bed?

We pushed through a set of double doors. Stainless steel and the rounded white edges of a dozen different monitors and instruments. A blaze of overhead lights. Gowned and masked people in every direction, most of them clustered near the pedestaled table in the center. All of them reduced to a pair of eyes. I felt Neva's arm stiffen. Then a dark-haired woman wearing a surgical cap festooned with circus animals bent low beside the gurney.

I hear you have a dog. Is that true? Because I love dogs. What's your dog's name?

Neva whispered *Coco* as the gurney approached the operating table and was slowly lifted so that the edges of each would meet. The woman looked at her again and spoke in a gently teasing voice.

I know you're a big girl, but I don't think you can go from that bed to this one by yourself.

So, of course, she did, and suddenly moments that had been crawling by for days screamed forward and the mask was on her face and her eyes glazed over and someone was saying *we'll take good care of her* as we were ushered back through the doors. The nurse's *the waiting room is to the left* echoed from miles away.

Stress takes many forms. At ten, I yanked open the door to our garage one night, believing the dog was inside, and came face-to-face with a bear. At fifteen, I was forced to drive a car I could scarcely shift through the chaotic nighttime streets of Guadalajara, its owner alternately laughing and vomiting in the back seat, my US learner's permit still fresh. I made a panicked wrong turn onto a one-way street, only to find lines of cars speeding my way, so I bounced the car over the curb and onto a grassy strip. Out of nowhere a Mexican cop appeared at the window and threatened to throw me in jail. At forty, I leaped from a lava cliff into a Hawaiian sea where the surge proved far stronger than I had judged. Diana was pale on the black rocks above as I fought for air and an exit. All of these moments were memorable, and yet still mild when compared to the horrifically endless list of acute stresses that have shaped the entire course of human existence.

Which is why we're not that bad at the immediate fight-or-flight stuff. We often manage to get our shit together reasonably well when the situation is: *You gotta deal right now*. As biologist Robert Sapolsky has noted, it's the chronic stresses that are the real insidious bastards. On that front, our brains are a bit of a mess.

Why aren't we better equipped? After all, many of the things that produce chronic stress are not unique to humans or even primates. All manner of creatures go about their daily lives with frequent fear of some kind, in most cases of being eaten, but it goes beyond that. The more we look, the more we see evidence that animals experience complex emotions of love and attachment, of longing, of grief for perceived or real loss. Is chronic stress a consequence of being sentient, of caring? Or a necessary by-product of the ways we dealt with charging lions and murderous neighbors? Do the forces of natural selection not weed out the slow erosion chronic stress can create? Are we just works in progress for whom the mitigation of such stress has been a luxury until only the very last chapter of human history?

Based on his extraordinary book *Behave*, I'm guessing Dr. Sapolsky might answer *well, yes* to all of the above. And also: *it depends*, before explaining that the particular assembly of elements that somehow fires our neurons and unleashes our hormones and shapes the way we think and act and feel is wondrously, hideously, astonishingly complex, and that each moment of our lives can be threaded from chemical firings just seconds before to eras that predate the existence of our species. And everything in between. He might also say:

Good God, man, go and get a beer before you stress out about it too much.

But as I wandered the hallways of the children's hospital or sat shifting endlessly upon the waiting room couch, all the while clutching my cell phone, waiting for the next report from the OR, I could not find a way to lessen the revolt within. Maybe I did something to stress out my four-year-old daughter without realizing it. Maybe I had given her cancer. Maybe I was giving myself cancer now. And if I was, good. Because I deserved it. All I could think about was failure: my failure, evolution's failure.

I shook out of that self-destructive loop only when the final call came, telling us she was out of the OR and where we could wait for the surgeons to give their reports.

Dr. Chan arrived first.

She was the smallest child I've ever done that for!

I assure you, those are not the first words you want to hear from a surgeon eleven hours after your daughter's brain operation began. But my panic eased when I saw the pride in his face and heard his buried lede.

She's doing fine.

Can we see her? Diana asked.

Soon. They are just getting her moved to ICU and settled; then they will call you in.

How did the surgery go?

All I can tell you is that my part went fine and she's doing well. Dr. Hankinson will be in soon to talk about the tumor resection.

But he glanced away as he said it, and another creeping tide of dread took hold. Soon Todd walked in and wasted no time.

Look, you guys, she did great. That's what matters most. But I want to tell you straight off that I had to leave some tumor behind. I know that's not what you—what any of us—hoped for. And most of it is out. But, unfortunately, a piece of the tumor was stuck to the optic nerve and some critical blood vessels. Taking that out risked damage to both.

Fight or flight takes shape in many ways. Sometimes you are overwhelmed with the impulse to bolt. Sometimes you fight back with everything you have. And sometimes, fight and flight both seem physically impossible and you simply feel paralyzed, unable to do anything. Physicists will tell you that gravity varies only a little bit depending on where you are—nowhere near enough that we would ever notice. Drop the apple from any tree a hundred times and the same simple equation will predict its fall with vanishingly little error. Yet there are times when Earth's mass suddenly seems to focus all its energy on you alone. Newton didn't have an answer for that one.

Still, as I sat pinned to the waiting room couch, I began to realize this—this crushing moment—was why we chose Todd. Why we put our faith in a man who, when confronted with that chance to say *To hell with the risks, I'm surgeon enough to pull this off* had the wisdom to choose no and the strength to deliver the hard news that may ultimately have saved our daughter's vision or her life. I remembered how his calm on the first day we met him had settled Diana and me down and how later we discussed his comfort with uncertainty. He didn't need to center his ego, and in that, he gave us confidence that he would see what the critical moments actually brought, rather than trying to force his own preconceived notions upon them. Unaccountably, I also began to wonder what particular little tweaks in our genes laid the foundation for a precious few to have the courage to both operate on a child's brain and to know when to stop.

In time I'd come to see how Todd was doing something else that science—if we let it—can cultivate in us all, perhaps helping us shore up that rattletrap factory of our minds. I had been caught up in cycles of self-directed rage and overwhelming stress, but much of that was all about my own ego. As those Costa Rican forests first reminded me, scientific excellence urges us to put our egos aside and just trust in the power of observation. True scientific practice is not about playing a role: the hero surgeon, the guilt-ridden father. It's not about imposing a simple story like Fate or Fault onto an infinitely complex world. Hopefully, once we are able to accept the fact that our stories don't always progress as we think and want and perhaps find ways to even wonder at those things we can't change, viewing the world through a scientific lens can also help us take in our unchangeable limits without drowning in shame or flailing in denial. Science is as much about maintaining boundaries as it is about taking risks.

It's about what Todd did in that operating room: staying present, patiently observing reality, and being open to adapting quickly in the face of observed data. Todd wasn't thinking about his reputation for excellence and bold decision making. He wasn't thinking about his reputation at all. He was thinking about Neva.

CHAPTER SIX

A few years before Neva was born, Diana and I sat on a plane bound for Hawaii—the land of my birth. It was our first big trip together as a couple. But she was pissed at me, and her stony silence let me know it. She had good reason. I was enmeshed in my painful divorce, and I had not done a good job of keeping its inevitable shrapnel from piercing our daily lives. Above all, I was locked into cycles of guilt around a life that would now take place far from my two young children. The oddness of my marriage had meant the kids had never known a time when we all lived together—from the time of their birth, I routinely traveled from Colorado to Massachusetts to see them. But the divorce would cement my role as a long-distance dad, probably for good.

It left me struggling mightily with feelings of failure. I'd grown up in a household with a model of a committed marriage and always thought I'd be good at it too. Yet here I was, walking away. The ramifications of it all too often made me a moody, despondent, self-indulgent, absolutely-not-very-fun pain in the ass. And Diana was over it. Shortly before the trip, she had cried and yelled and then dumped me, shaking her head as she walked out my front door.

She walked back through it late that night, still mad, still uncertain.

I'll go on the trip, she said. *But after that, I don't know.*

There's a particular kind of agonizing force field that envelops you when you're trapped for hours on a plane next to someone who you love, who is giving off vibes they may not love you back. I responded by talking too much. For a while I tried to appeal to her love of science by weaving together some of my own history in Hawaii with stories about how the place came to be.

I told her about a December day when I stood before a coralline sea and threw a fit. I would. Not. Get. In the boat. In my defense, I was only four years old—the same age as Neva at her diagnosis. The small outboard contained my parents and a couple of their friends, and it was bound for Chinaman's Hat Island, located just off the northeast shore of Oahu. Although Hawaii was the only home I knew, I wanted no part of leaving dry land. They could go for all I cared; I was staying on the beach. Then, after losing the battle, I became transfixed by watercolor images of the shallow ocean gliding by, prismatic blues and greens so all-consuming that when we did reach the island, I threw another fit about getting out. It's one of my earliest memories of pure astonishment at the beauty of the natural world.

Then I said:

Now I know the crazy thing is that what really entrances those of us who love Hawaii is the beauty of death.

That got her attention.

What?

Well, it's the volcanoes. They destroy everything, but that's what makes the whole thing possible.

And from there I launched into a story about how I taught one part of my introductory ecology class. (Yes, I was still talking too much.) I would ask the students:

What sustains life?

One year, after the predictable moment of nervous silence, the answers went like this.

Oxygen?

Water?

Sunlight?

Beer!

We all laughed. Then I showed a picture of a wheelbarrow full of dirt and began to talk about how the world's most fertile soils all start in disaster. Pyroclastic explosions of ash and lava slam into hillsides and streams, obliterating trees and boiling fish alive in the water. Or massive glaciers slowly pulverize everything in their path... then unleash a catastrophic flood for good measure. The aftermath is a horror—a moonscape of ruin. It is also a beginning. Flooding rivers and falling ash leave behind a bounty of minerals, which in turn pave the way for life. It's no accident that the world's major early civilizations emerged in river valleys. Tigris, Euphrates, Yellow, Indus. Not only could people get water easily, but they could grow crops in deep and rich soils fueled by past floods. Fast-forward to today, and again you find Earth's breadbaskets in places like the vast river plains of the central United States and northeast China. These places grew good dirt first, meaning they can grow lots of food today.

Diana didn't laugh at the whole *Beer!* thing. She just said, distantly:

Uh-huh. I know all that stuff. What does it have to do with our trip?

I swallowed and pressed on. We were headed for the Big Island, where a tapestry of destruction and renewal is there for everyone to see. I talked to her about how in some places, you could still walk across blankets of ash, while right next door, a thriving forest obscured the eruptions that had come only a few years before. In other places, the clock just takes a bit longer. The dried lava you can see in many parts of the Big Island is still a fresh start on which something finds purchase. A hardy plant takes hold in a tiny pocket of dust in the lava, creating the first splashes of green on a stony backdrop. Or perhaps it's a coating of gray or green or even orange lichens—partnerships between algae and fungi—which call the new land home. Each outpost of life adds concentrated carbon dioxide to the water already flowing through the rocky cracks, fueling chemical weathering and the release of mineral nutrients therein. Soon enough, the dead world becomes effervescent and green.

Embarrassingly, I think I actually used the phrase "effervescent and green" at the end of my mini-lecture. Diana remained unimpressed by this particular brand of poetry. Her only reply?

Thanks, Professor.

So I squirmed and suffered as the minutes crawled by. But as the plane finally brought the island into view, she couldn't help but soften at her first sight of Hawaii. Her face was glued to the window as she reached back and touched my leg.

Wow, look at that!

I leaned over her, now able to see the meeting of lava and sea along the Kona Coast. The surface below was ropy and gleaming black, interspersed with pale bunchgrasses wedged into the rocky fissures. It surprises some visitors to learn that this stark landscape was once the preferred home of Hawaiian royalty and thus bore witness to a rich and sometimes bloody history. As chronicled in Gavin Daws's terrific history of Hawaii *Shoal of Time*, it is where Kamehameha I took up residence after he defeated his enemies and launched his attacks to unify the islands and where the *kapu* system—a harsh set of laws derived from spiritual beliefs that, when broken, usually meant death—occasionally unfolded in brutally violent ways. As I continued to worry about where Diana stood, I hoped the land would be kinder to me now.

The massive bulk of Mauna Loa was to the right, the rougher and smaller outline of Hualalai Volcano riding upon its northwestern flanks. Just beyond, you could see snow glistening on the summit of Mauna Kea. When we dropped into our final approach toward the Kona airport, I took a chance and pointed at a jagged stretch of coastline nearby.

There's a pretty special spot hidden in all those lava rocks that should be the first place you jump into the Pacific.

She turned and answered with a slight smile.

Can we go there right away?

Phew.

Soon enough we were in swimsuits, picking our way across a slippery bench of lava as waves crashed against its edge. She paused and looked at the violent collisions.

Exactly where do you expect us to jump in?

The skeptical look that Neva inherited was on her face.

It's just ahead and back from that edge. There's a system of old lava tubes below us—places where the lava ran to the sea and then shifted to somewhere else, basically leaving tunnels behind. And one of those ends in a hole at the surface, which fills with seawater.

I'm gonna jump into a lava hole.

Stated flatly, with the same look.

Well...yeah. But it's big, like a big Jacuzzi, and the wave pressure is going to make the water go up and down, uh, maybe kind of a lot. It's going to look a little crazy, like it will suck you out to sea through the tunnels. But I promise it won't! It's fun!

A long stare.

We reached the edge to see the water come rushing into the hole, spilling over the top and soaking our feet. Then it all reversed as the tendril of ocean was sucked back toward its home, quickly dropping the water level a good six feet or more below. The hole did not go dry, but for a moment it looked as though it would. And it appeared certifiably insane. I turned to her, ready to say, *We can pick another spot.*

She jumped in.

She laughed as the pressure shot her back toward the surface, swirled her in foam for a moment; then it all went rushing downward once more. She treaded water more vigorously in a brief moment of alarm. Then, when she realized that she would not be pulled into a dark and watery grave, she floated on her back and laughed again as the next wave brought her skyward. I just stood for a moment and watched as she took a few more rides. That smile, that giant, radiant, always a little shit-eating grin. When I jumped in too, she grabbed my hand but didn't speak as we rode not only the water but a moment in time. Something that perhaps science could explain, but why would I ever want it to?

We stayed for an hour or more, letting the ocean wash away the tension of the flight, then drove to the other side of the island. We were headed for the little town of Volcano, whose scattering of small buildings was nestled amid red-flowered ohia trees and prehistoric ferns with giant fronds that could reach well above your head. Each of which grew nowhere else on Earth. By the time we arrived, the warmth of the Kona sun was long gone, replaced by a soft mist and a vague smell of sulfur. The still-active Kilauea volcano was right next door.

Hints of the conflict surrounding my divorce had emerged once more on the drive, but for the most part, we managed to set it aside. We stopped at a Hilo grocery store on our way, stocking up for the week and including a couple of to-go containers of tuna poke, as common and varied behind Hawaiian deli counters as sliced meat and pasta salad are anywhere else. Now we sat on the porch of our small rental cabin in the fading light, scooping up mouthfuls of the fish with disposable chopsticks. Diana, a near-pathological runner who would end up running all but two days of her pregnancy with Neva, wanted to know where we could go in the morning. That one was easy.

My favorite run on the planet starts just down the road. Goes from rainforest to open volcanic crater and back. Amazing.

How long is it?

The important thing.

I think a bit over ten miles if we do the whole loop.

You sure you can handle that?

I was not the runner she was, and she gave me a devilish grin paired with a hint of challenge in her eyes. I replied with false bravado, knowing that she could well be right. I might not be able to stay with her, a thought that brought a stab of pain as I hoped that only meant the run.

The trail we would follow was full of signs of death and destruction, in the wake of Kilauea's recent eruptions. But some places would have signals of tenuous hope for what might come next. A mirror for what I felt inside. A small island of trees that somehow escaped the eruption's wrath. Or a scattering of tiny new saplings trying to find their way in a sea of lava and ash. Some of those trees would make it and by doing so open the door to more. And in other parts of the run, we'd see the end result: a reborn forest that was full of new life and that in a few short decades had completely hidden its dystopian birth.

We began the loop by jogging down our short side street, turned left past the post office and its bordering hedge of bright purple tibouchina flowers, then crossed the main highway and entered a tunnel of trees. The trail wove through the ohias and ferns for a few miles, a forest so otherworldly that it felt as though a dinosaur could be around the next corner. We ran largely in silence, in part because I needed to save my breath to keep pace. The first time I'd run with her along the Osa's coastline, she started down the trail with a slightly stiff and awkward gait, and I thought:

Hmm, she doesn't seem like much of a runner.

In time I'd see others make the same mistake, followed by her leaving nearly all of them behind.

Eventually we broke free of the forest and onto a section of the trail that crossed an undulating landscape of gray ash sprouting ghostly remnants of seared trees. Then we dropped down to the floor of Kilauea Caldera itself, our path visible as a slightly lighter shade on the black pahoehoe, before climbing up past the crater observatory. As we turned back toward the cabin, she slowed and then stopped on a pile of lava that served as a tomb for the forest it once destroyed. The wind whipped, and she kept brushing her dark hair back from her flashing eyes. Then she let out something between a yell and a gutting plea.

How can you ask me to believe you love me when you still allow so much space for the bullshit around your divorce?

She turned away and took off down the trail, leaving me gasping to catch up.

That night I found the courage to stop trying to grandstand or pontificate and just be real with Diana. I choked out my feelings of failure. I couldn't really explain why I was making some of the choices I was, I told her. I just was trying my best to navigate a situation I never expected and that broke parts of my heart. Then I said I loved her so much that the hours after she'd left were among the worst in my life. And if she needed to go, it would crush me but I would understand.

She looked at me without speaking for a long time. Then she walked out on the deck of the cabin, ignoring a light rain as she sat for nearly an hour in a bloodred Adirondack chair. When she returned, her eyes were brimming and bright and she simply said:

Okay. I'm here. Now what?

Then she wrapped her arms around me and held tight as we rocked slightly back and forth and listened to the rain.

The moment returned as we held each other in that white-washed hospital room after Todd left. In Hawaii my walls had broken down, and when I let them be destroyed so that a new story could emerge, our relationship had metamorphosed beyond simple attraction and into an unbreakable bond. I thought of how the vulnerability of those moments and the fear of a situation that was teetering on the edge of destruction unaccountably led to bright new flowers sprouting in the debris. Just like the landscapes that surrounded us. Just like the world we all share. Science shows us that nearly every seeming apocalypse is ultimately ephemeral and that many of them are followed by unimaginable beauty.

CHAPTER SEVEN

Neva woke up from her surgery much more composed than her parents. Framed by tangled hair and a swollen face, her eyes were still bright and searching.

Mama, she said, matter-of-factly. *When do I get my ice cream.*

It was a statement, not a question. Then she sought out the stuffed giraffe that had shared her bed since the week of her birth and began to assess her surroundings: the blinking monitors, the tubes and coated wires running from her chest and arms, the whiteboard above the foot of her bed. She asked, *Can I draw on that?*

It was a fair question from a four-year-old, who thankfully could not understand the cycles of anxiety that the whiteboard conveyed. Like most such boards in any hospital room, it listed a few basics: Neva's name, that of the nurse on duty, a series of medications. But in this case, the nurse would also put a new number up every couple of hours: the concentration of sodium in Neva's blood.

Yes, salt, more or less. We all crave it for a reason: sodium concentrations are critical in regulating a whole host of different cellular functions. Perhaps you vaguely remember talk of the sodium-potassium pump from your high school biology class? Well, that thing is a big deal throughout our bodies and especially in our brains, where something like half of its basic metabolism is linked directly to one of these little pumps.

And it's not just us. Think of how deer or cows or birds will converge on a salt lick. Yet we take sodium somewhat for granted—I mean, it's just *salt*, for godsakes—even those of us whose job it is to study the elements. In my field of biogeochemistry, you could probably coat the entire planet in the pages of articles written about how nitrogen and phosphorus are critical for life, how they shape the ecosystems on which we all depend, how the discovery by humans of how to turn them each into countless bags of fertilizer is one of the most significant turning points in history. And it's all true. But meanwhile, everybody needs their sodium, something University of Oklahoma ecologist Mike Kaspari began to think about years ago. First, from the perspective of an ant.

Kaspari and colleagues discovered that ants in tropical forests like the kind I studied in Costa Rica were not, as much as it might seem to me otherwise, constructing their daily schedules around where I sat down for my lunch but partly around how much sodium was present. How did the scientists figure this out? By doing a version of what you might do if you found ants in your home: they put out ant bait, but of different kinds. Some contained sugar, some salt. And as it turned out, in places where sodium was in shorter supply, the ants came running for the salt traps. Just like those deer, cows, and birds.

Put another way, the amount of salt hanging about could predict what ants you might find and what they were doing. Scientific papers are not often celebrated for their gripping narratives, but every once in a while you pick one up that makes you go: *Goddamn, that is cool!* For me, Kaspari's work was one of those.

But now sodium was no longer cool at all. Every two hours, a nurse would take another aliquot of Neva's blood and ship it downstairs for a measurement. It was the only reason she had to stay in the ICU beyond the first thirty-six hours. Her surgery often blew apart the body's ability to properly concentrate water, because the pituitary gland releases hormones that are essential in regulating sodium concentrations in our bodies. And when that went wrong in a healing kid, the ensuing cascade could get dangerous in a hurry.

On the first night after her surgery, Neva seemed remarkably unfazed by the blood withdrawals. The nurse could take them from her A-line, so no new needles were required. Our daughter woke for a couple of them, slept through others, and seemed more herself than we expected after what she'd just been through.

It didn't last. On the morning after surgery, a team of residents and medical students surrounded her bed as she still slept, then woke her before we could intervene. The resident in charge gave a half-hearted *How are you this morning?* before making a show of lifting one of her arms and launching into a summary of her condition and status. Neva's eyes widened in fear and confusion.

Mama? Am I getting a shot?

Diana grasped her hand with a *no, honey* before turning on the resident.

What the hell do you think you're doing?

Excuse me?

You heard me.

His answer was haughty.

Ma'am, we're doing morning rounds.

No, you're scaring the hell out of my daughter for no reason at all.

Well, ma'am, I hardly think that—

Get out. Right now. Wait for us in the hall, and we will be out in a minute.

The resident was taken aback and looked ready to retort, before thinking better of it and leading the group out the glass doors. Though he could not resist a *we can't wait long* as he walked out. Diana had a look I knew well—it meant trouble was coming—but she masked it as she turned to Neva again.

Honey, don't worry. That's just the doctors checking on everyone. We are going to go out and talk to them. You'll be able to see us the whole time.

Mama, why are you mad?

Oh, I just didn't like that they woke you up without asking us. But it's okay. You rest and we will be right back. Is that okay?

You'll be right where I can see?

Diana pointed to the hallway through the glass.

Yes, honey, right there.

Neva gave a small nod.

We walked the few steps into the hall and slid the door closed. At maybe six five, the neurosurgery resident was nearly a foot taller than Diana, but she walked to within inches of him while fixing him with a venomous stare.

You're a pediatric resident, right?

The man looked hesitant, not expecting this to be her opener.

Yes, neurosurgery.

So why is it that you don't seem to have the first goddamn idea how to treat a kid who's going through something terrifying? Did you fall asleep in all of the trauma lectures?

Ma'am, I don't think this is a productive way to—

Don't you dare tell me how to talk to each other after you just came in and woke up my child for no reason and then scared her half to death.

She wasn't done. She even jabbed a finger into the tall man's chest.

Here's how it's going to work from now on. You will not enter the room without checking with one of us first unless it's an emergency. You will not talk about her tumor and her wounds and all the rest of the shit you said in there in front of her. Ever. If you break those rules, I'm getting Todd on the phone and doing everything I can to get your ass booted out of here.

The medical students' eyes widened. The resident flushed but said nothing in reply. Diana went on.

Now, what time this morning is her post-op MRI?

We scheduled it for three this afternoon.

My wife paused in astonishment. When she spoke, the words came out slowly, laden with barely contained rage.

It's seven a.m. You mean to tell me that after all she went through yesterday, now she can't eat or drink for the next eight hours? Are you fucking sociopathic, or just completely incompetent? Get with radiology right now and move it up.

The resident's answer came out as an apologetic squeak.

I—I don't think that's possible.

Make it possible.

But it wasn't, and it pushed Neva over the edge. By nine she was crying for water, sobbing harder as the morning wore on. By lunchtime she began to withdraw, her eyes dull once more. By the time the MRI prep nurse finally arrived, she no longer spoke and would not even put the lip balm on her mask. She remained silent unless it was a fearful question about getting poked with another needle.

Which began to happen frequently. The A-line failed, as did the backup IV used to draw blood, so they began to stick her finger around the clock, in large part because of how hard it was to get IVs into her. So they slowly squeezed drop upon drop of blood, the tiny tube agonizingly slow to fill each time.

Meanwhile, we demanded and charted every number, hoping for the stability in her sodium numbers that would send her off the ward. And we lay with her, talked to her, read and sang to her, fighting back tears as she refused to say a word. As she looked away.

It went on for days, each of them excruciating in both the unbreakable shell of her withdrawal and in the regular blood draws that would make her entire body go tense. Only then would she cry out again and again:

I don't want this!

The child psychologist made repeated visits with a battery of toys and stories, while the nurses tried to reassure us that they had seen kids react this way many times before. That it was normal and would not last. That did little to settle Diana, who still fumed about the arrogant resident and the botched MRI. At one point she ranted for several minutes about the need to pay better attention to the mental state of the patients in the ICU; then she stalked out of the room. I felt a moment of pity for any other wayward resident she might encounter.

Weeks before her surgery, I had stood with Neva in our kitchen and pulled out a bright silver bowl, a box of cornstarch, a cup of water, and a vial of purple food coloring. I helped her mix the water and cornstarch and then add a few drops of color. She stirred it all happily, pointing out the swirls of purple as the color took hold.

What are we making, Dada?

It's called Oobleck, honey.

She laughed.

That's a funny word!

But her expression turned skeptical when I told her to yell at the bowl.

Why, Dada? That's weird.

Just try it.

She let out a half-hearted *WOOO!* and then looked at me quizzically.

Try it again. But louder!

She let loose enough to make the dog bark, and then her eyes widened in surprise when sound waves appeared on the purple surface.

Oobleck is a non-Newtonian fluid, and that means it doesn't follow Isaac's simple rules for how we think the world should work. Treat it gently and it will flow like water; smack it around, and it hardens into a solid. The elements within run for each other under pressure, as though mobilizing into some unseen defense strategy, suddenly creating a wall to the outside world. And in that, Oobleck—like the caterpillar—has something to teach us about ourselves, for it has the plasticity of the human mind and many of the same limits. We don't do well as scientists, or as human beings, when we allow stress and fear to stand in our way. There's physics behind this—the very same physics that produced the sound waves in Neva's purple bowl.

Believe it or not, our blood becomes more viscous when we're stressed, as part of a process called hemoconcentration. We've known about the dangers of this phenomenon for a long time: you can find medical journal articles about the links between hemoconcentration and shock and trauma that emerged from studies of soldiers in World War I. It can happen for a number of reasons, but at the core of it, it means plasma, the carrier of the body's blood cells and waste products, drains away; concentrations of protein and hemoglobin go up. This is thought to be one reason chronically stressed people deal with much higher rates of heart attack and high blood pressure: not diet, not "lifestyle," but the sheer fact of our non-Newtonian Oobleck nature. Anguish makes us denser and less fluid. Literally.

All three of us were trapped in our own forms of such distress in those first days after surgery. There was nothing fluid about us. Neva remained silent and fearful, Diana swung from almost desperate attempts to break through to our daughter into moments of barely controlled rage, and I felt like there was nothing helpful I could do. I was learning firsthand how in the pediatric ICU, everything becomes non-Newtonian: a bad moment quickly hardens into something terrible; a moment of relief flows through and over you like a warm bath. Diana and I remained laser focused on those sodium numbers because we knew they were the ticket out of the cycle we were all in.

And then, near the end of the first week, Diana made the choice to simply climb in bed beside our daughter and draw a picture of our home. She sat quietly beside Neva and drew the outline of the house, the back fence, the deck that was outside our living room. She told Neva what she was doing in each step, asked her questions about what she should draw next, and continued to form the picture and move to its next element when Neva would not answer. As she did, I saw a hint of a spark return to my daughter's eyes and a flush of new color on her face. Then she broke her days of silence.

Mama, she said. *You forgot to draw Coco.*

From there, she began to return to herself, and before long we were out. When her final sodium checks all came back in range, unaccountably I pictured Kaspari's ants and told Diana as much. She gave me a baffled look. But it was soon forgotten when the nurse told us we were free to go home. We paused for a few seconds and breathed deeply after we walked through the big glass doors and blinked up at a cold but clear December morning sky. Neva was perched just behind in a red wagon laden with blankets, where she was nearly hidden beneath a giant stuffed lion that a friend had brought to her room. It was three days until Christmas.

Before long, she sat between us in a booth at a Denver sushi joint, cracking up the waitstaff by happily eating pieces of raw fish and handfuls of salmon eggs. Her hair was still an impenetrable tangle and her hands were bandaged, but her eyes were bright as she talked about anything. Everything. After a while, her chatter turned to bicycles.

Dada, when will I get a bike? I want a bike.

Honey, you can have a bike today.

We found a tiny white one with a floral basket, which she rode around the shop's rows of much larger models. By dinnertime, we rearranged the living room so she could do laps around the couches and entertainment center. Every so often she hopped off to give the dog a headlock and get a face wash in return. That night she fell asleep beside the dog, happily mumbling about what Santa might bring.

Can we make great discoveries, can we break through to something remarkable—in science or ourselves—when we are hardened like Oobleck into overly reactive balls of hemoconcentrated stress? Sure, it happens. But true learning, discovery, and genius are far more likely when we're in a playful state of mind, relaxed enough to be open and observant and accepting of what both is and could be. A host of psychology experiments show that people are better at solving problems and breaking through with new insights when they are put in situations that relax their brains. The same goes for retaining information—we lay down lasting memories more effectively if our brains are not a hardened mess. And the behavioral data are backed up by neural imaging: stress us out too much, and some critical areas of the brain for creative insight essentially shut down. In *Behave*, Sapolsky describes one part of the cascade of barriers our brains put up as a response to stress in typically evocative style:

> Stress compromises frontal cortical function. Glucocorticoids accomplish this by enhancing norepinephrine signaling in the pre-frontal cortex so much that, instead of causing aroused focus, it induces chicken-with-its-head-cut-off cognitive tumult.

So, as the cliché goes, we need to chill out. Good things will come from it. But that's easier said than done, of course. We are pretty hardwired to slip into fight-or-flight mode under stress, which is no accident, because for years our survival depended on it. Yet science not only helps us understand this reaction, but it also shows us that we can trick our own brains out of it. Making active choices to alter those external stimuli in whatever way possible, be that a certain kind of breathing in the face of danger, shifting our attention to something more peaceful, getting the hell out of there and doing something better, or simply turning off that smartphone, all reroutes the internal supply chains in our brains to restock the creativity shelves once more. That's what Diana did when she settled herself down and began to draw beside our daughter and what Neva did when she jumped off her new bike to tackle the dog.

CHAPTER EIGHT

Diana could lock down and escape into her science like nobody I'd ever known. She'd be there but not, consumed by the laptop screen or the notebook before her. Sometimes it annoyed me more than I'd admit. Once Neva was born, we had a running joke about if I'd make her top five. Neva, the dog, her coffee maker, and her science left me clinging to a spot on the list. We'd laugh about it with friends, but when she disappeared again and again into her unseen bacterial world, the joke wasn't always funny. And yet, as the years passed, I realized that her passion gave her an anchor that let her show up for others in extraordinary ways.

I saw hints of it on that first trip to Costa Rica. She was out of the country for the first time and surrounded by a diverse menu of potential stressors. There were language barriers and venomous snakes, and scorpions in the room where she slept. But she couldn't stop talking about bacteria. How cool their world was, how much we didn't know. She had joined the trip to try to answer what sounded like a basic question: Did the identities of the bacteria in any given soil community help determine how fertile that soil was?

The problem is, those soil communities are mind-numbingly complex, something we didn't know until fairly recently. Prior to the 1970s, microbiology was all about microscopes and lab cultures. But scientists couldn't readily see a lot of what's out there and couldn't make it grow in the lab. Then National Medal of Science winner Carl Woese began to look at the microbial world via its DNA signatures, not its appearances. A couple of decades later, microbiologist Norm Pace figured out how to take that approach into the natural world in a big way and created a new tree of life, which revealed just how astonishingly rich this unseen world is. A single spoonful of that Costa Rican soil will contain perhaps a trillion bacteria, and the genetic diversity of bacteria on earth dwarfs what exists across every animal and plant we know. Stink bugs and hummingbirds, lions and humans, seaweed and giant redwood trees, and everything else occupies an embarrassingly small corner of the tree of life when held up against single-cell life forms. So, figuring out who was in that soil, what they did, and how it all might connect was an extraordinarily difficult task.

Which is why Diana loved it. I remember her excitement as she talked about it all on one of those coastal runs, me just trying to say a few words in between my heavy breaths, her talking endlessly almost as though the science fueled her lungs.

That run was the first time I began to appreciate what was truly in—and on—us. She said, *Do you know that something like half your cells are not yours?*

Pant, pant. *What?*

Yeah! They are the bacteria in you. And on you. Think about everything they probably do!

Can they help me run any faster?

She was right to be fascinated by it. As Ed Yong chronicles in his book *I Contain Multitudes*, the years since have shown us that we aren't just a passive transport vessel for bacteria. I like this passage from Yong's book:

> Every one of us is a zoo in our own right—a colony enclosed within a single body. A multi-species collective. An entire world.

And this one, which hints at the ways our worlds intersect with others via the zoos we carry:

> Within 24 hours of moving into a new place we overwrite it with our own microbes, turning it into a reflection of ourselves.

Those microbes might not grant my wish of running better, but many of them are essential to our health. They fight off disease, help digest our food, produce vitamins, maybe even determine how well we sleep. When these partnerships fail, we suffer. We're also starting to learn that a host of autoimmune disorders—diabetes, some forms of arthritis, multiple sclerosis, and more—may be passed down in families not via our genes but via a faulty set of microscopic partners.

A few of those partners are even *in* our genes. Over the course of human evolution, some genes have entered our DNA not from our parents but from a host of other life forms. And no, this isn't something out of a bad sci-fi movie; it happens by a process known as horizontal gene transfer. A bit of DNA floating out there in the world makes it into a microorganism of some kind—bacteria, viruses, or tiny single-celled beings known as archaea—and from there, it latches on to and in to us. We live in a deeply interdependent world, including within our own bodies, and a remarkable amount of that interdependence rests upon the activity of these life forms we typically cannot even see.

Diana loved it all, and it became the organizing quest of her professional life. What was out there? How much did that matter to our lives and to the functioning of the natural world? She wanted some answers, yes, but what stood out was her unrelenting joy at the very complexity and uncertainty of it all. Hardly a day would pass when a new question wouldn't pop up, but she didn't seem to care that many of them might be unanswerable, at least anytime soon. If ever. She just loved to think about it all and pick away at the mysteries.

Prior to meeting Diana, my relationship with science was, well, less admirable or nuanced. I used to walk around with near certainty that given enough time and effort, science could explain pretty much everything in our lives. I thought that was its greatest purpose. And with that came a belief in a moral and intellectual superiority that I kept inside…but, in truth, felt.

Decades ago, David Quammen wrote a piece contending that crows are too smart for their own good. That it makes them bored, from which their reputation as one of the jerks of the animal world is born. They are smart enough to knock out their daily tasks with great efficiency, then sit around preening and inserting themselves into the affairs of others with a distinct air of superiority. But they aren't smart enough to recognize their flaws. Neither was I.

My ego helped destroy my first marriage. For years I wanted to blame my ex for nearly all of our problems. She had her issues, sure, but so did I, and some of those bubbled up from a toxic wellspring of scientific absolutism. At one point, as we were planning our wedding, we fought about a church service. She wanted us to be married in the same quaint and white-planked Cape Cod chapel that her parents had been married in, but I was too much of an atheist asshole not to push back on that plan.

For the younger me, science disproved the existence of God or any other forms of mystical belief. I followed my father, a former physics major and a philosopher of science, who I once heard say:

Evidence is the only God I need.

I found claims to the contrary to be soft and denialist thinking, in which one could find the roots of nearly every societal ill. So, while I fooled myself by saying the wonders of the natural world gave me a kind of spirituality—as they eventually would—the truth at the time was that I just thought they were cool.

Today I realize how my arrogance was another form of zealotry, inconsistent with the very foundations of what a scientific mindset should be. Science, like religion, has extremist sects: eugenics, race "science," quack medicine. For all its miracles, science also harbors violence and exclusion, racism and hate. Its history is as sordid as any, an ugliness that stems directly from human flaws, not something fundamentally wrong with many of the basic tenets of the pursuit. It is, in so many ways, eerily similar to the difference between the ideals and practices of many religions. Christianity exhorts its followers to love thy neighbor as they love themselves; then it is used to exclude, demean, and even kill. But so is science. We scientists are supposed to be unbiased, driven by the purest forms of curiosity, ready to question and admit we are wrong. But we're humans. We fall short.

And yet I've come to believe that only through centering our humanity in all of its messiness and wonder does science reach its best forms. Science, like Christianity or any other faith tradition, is often not about a fixed reality. It's a human construct of its own, a way of seeing the world. And that world is constantly in flux, far too complex and unpredictable for us to ever fully pin down. We can—and should!—keep digging into its mysteries. But in doing so, we have to retain the humility that comes with understanding our limitations. The very essence of science is to question what we think to be true. And therefore we must question the limitations of science itself.

For me now, what C. S. Lewis implied about Christ is also applicable to science: it's a light by which we may see all things, including our own frailty, with wonder and acceptance. It is a means of dancing beautifully within our own limitations, in ways that occasionally redefine what we think those limitations are. No more, and no less, than that.

Somehow, Diana seemed to know all this innately. Her escape into science was not a separation from her humanity; it was an expression of it. It was there she felt most pure and alive, and she carried that over to how she practiced science. She loved any idea, no matter how crazy. She deliberately sought out and lifted up students who did not fit traditional molds; she pushed back against a focus on "classic" credentials that in turn becomes a path to racist, sexist, and classist gatekeeping. For her, science truly was an outlet for her love and her faith.

I first saw this clearly on the heels of Hurricane Katrina, after it devastated her adopted home of New Orleans. I came home to find her glued to the TV, crying. She stared at image after image of broken levees, of flooded neighborhoods, of death and poverty and displacement, while she cursed the president and FEMA head for their bungled response. When a helicopter panned over the wreckage in the Ninth Ward, the place where she spent day after day painting walls, restoring a park, learning a neighborhood, she knew she had to do something to help.

We finally made it to New Orleans more than a year after the hurricane, driving straight from the airport onto eastbound Saint Claude until we reached the lower Ninth Ward. Here, we found house after house still branded with spray-painted *X*'s, many of them bearing numbers in one quadrant that let you know how many bodies were found inside.

We reached a neighborhood that was quite literally gone, where Diana walked slowly from the car on an uncharacteristically frigid January day as her breath floated before her face. A shell of a house with broken walls of stained yellow bricks and a partially collapsed roof stood to her right. It was the only remaining structure on the block. An orange *X* was beside one darkened and paneless window, with a heart-wrenching 2 in one lower corner. She paused and stared at the remnants of the house for a long time. Then she spoke, slowly.

The park was right over there. The one we restored. At least I think so. I don't even recognize this place anymore.

When she spoke again, it was with quiet force.

I'm changing my teaching. We wave our hands all the damn time about a more sustainable world. This is the real thing, right here. This is what my students need to see and understand.

By the next fall, her course on Katrina was a reality. She dove into every aspect of the science of New Orleans that she could find: not just the role of her bacteria, but the social science of poverty and policy and greed, the hydrology and engineering of a city fighting a losing battle with water, the climate science behind hurricanes, the public health implications of overrun levees. When she learned former FEMA head Michael Brown now lived just down the road from us, somehow she dug up his number and called him.

How did that go? I asked skeptically.

He's coming to the class. He was pretty cool about it, actually.

You're gonna have THAT guy in your class?

She gave me a look of mild reproach and disappointment.

Isn't that the whole point of what we are supposed to do? Challenge people to think beyond whatever story we think we know? Challenge ourselves?

I stood shamed as I realized this was just her, once again, practicing what a scientist should. She wanted to know the real answers, not the ones everyone simply assumed were true. She wanted her students to have the same chance. And she was comfortable with the fact that some answers might not be satisfying or might not come at all. The point was to try. The point was to be both human and rigorous all at once.

Now, as we faced the realities of Neva's future, we needed all of that humanity and rigor. She'd made it through surgery, but since there was tumor left behind, we had another gutting decision to confront: Should we radiate that remnant mass or not? Diana immediately buried herself in the scientific literature and began to reach out to specialists around the world.

A picture emerged, one in which the perils of radiation were most pronounced in kids below seven. Neva was not yet five. Still, we found radiation oncologists who downplayed those risks, insisting they could minimize the collateral damage and that the benefits of stopping the tumor in its tracks were paramount. Other oncologists preached caution with stories of learning deficits and permanent blood vessel damage and other cancers to come. Digging through the scientific literature led us to a man in Germany who oversaw the longest-running study of craniopharyngiomas in existence. Diana wrote to him.

Dear Hermann, she typed. *What would you do?*

He wrote back. *Wait.* And though not yet published, he shared the data on why. As long as they were watched carefully, no greater tumor control was gained in radiating kids out of the gate versus waiting until signs of growth returned.

Statistically, the data could only speak to averages. They could not distinguish between a kid whose tumor remnant was a comfortable distance from the optic nerve or hypothalamus and a kid for whom the remnant mass was knocking on the door. Neva fell into the latter category. We pushed back on Hermann, sending him the post-surgical MRI images and explaining our fears. *Are you sure?* Back and forth. Eventually he gave us a bottom line.

Based on our study and what I've seen from radiating young kids, if this was my daughter, I'd at least wait until the two-month MRI. Then make a decision.

Eight anxious weeks later, Neva sat on Diana's lap, still a little groggy from the anesthesia, as we waited for those results. I stared at the same pictures and notes on the wall. The same memorial to the little boy with the baseball bat. I began to make trip after trip to the water fountain just as an excuse to move and not look at the heart-wrenching board.

I felt my faith in science being tested as never before. Were we stupid to skip radiation based on Hermann's data? Had we succumbed to the influence of our own profession and let averages and trends guide us more than knowledge of our own daughter and the specifics of her scans? What if the tumor now enveloped the optic nerve? A faith in the power of science is easy when you're dealing in generalities. It can be a lot harder when that faith must be pointed at someone you love.

Finally, Todd walked in.

Hi, kiddo. Do you want to see a picture of your brain?

Neva nodded slightly as I tried not to make too much of his upbeat tone. He turned to us.

Things look great. We got a nice surprise. Have a look.

I floated as the images began to flash across the screen, digital slices of her brain rendered in black-and-white. Sometimes their orientation was obvious — there's a jaw, nose, eye sockets — and other times I was guessing. Todd sped among them before settling on a cross section and zooming in on a region above her mouth. He pointed to a crescent border between black and shades of gray.

Remember, this is where I had to leave some tumor. If I pull up the scan from right after her surgery—he paused to hit a few keys and then dragged a similar cross section beside the first picture—*you can see the tumor remnant here.*

I looked at the kidney-shaped object on the December scan and then back at the new one. The spot was an empty black. I had a moment of soaring, illogical hope.

Where is it? It can't be gone, right?

No, it's not gone. But it fell away from the optic nerve and is now resting down here—another point with his index finger—*on the pituitary. This happens sometimes.*

Oh my God. That's good, right? Has it grown?

Yes, it's very good, and no, it has not grown. Even better, there is now a significant gap between it and the places we want to avoid.

That night we celebrated with ice cream for Neva and a bottle of wine on the deck for us after she fell asleep, as ever, tucked beside the dog. We knew this was an astonishing reprieve, so we soaked in a level of relief we had not felt in months. But it was far from a settled story. The tumor could easily grow again. The fact that we'd lucked out in round one didn't really prove anything yet. And so, the next morning, Diana greeted me with an announcement.

I'm going to Hermann's meeting in Germany.

What?

The cranio conference. He holds it every two years. Many of the best cranio docs and researchers in the world attend.

Wait, how do you know this? How can you go? Is it an open meeting?

No. It's by invitation only. But I emailed him and asked if I could give a talk from the dual perspective of a biologist and a parent of a cranio patient. He said yes.

Looking at my wife, I realized none of this truly surprised me. Of course she'd done this. Less than a month after her announcement, she flew to Amsterdam, then boarded a train for the German village where the meeting would be held. She'd go on to eat and drink and laugh with specialists from multiple countries, to challenge some of their claims in open discussion, to win many of them over despite her outsider status when it became clear she had done her homework. And she came home with newfound resolve that our wait-and-see approach was the right choice but not because the evidence was overwhelmingly clear.

I began to realize that her peace with the decision came because she'd left no stone unturned but also because she was innately comfortable with the limits of science. She said as much one night, turning philosophical as we sifted through point after counterpoint on the treatment of our daughter's tumor.

Too many people think science should give them certain answers. That's our fault.

I tried to pivot from the specifics of our decision.

What?

I mean, everyone jokes about how weather forecasts suck, when in truth they're incredibly good but won't ever get all the details right. And we keep hearing about a cure for cancer. There's not going to be a total cure. Science isn't going to solve every problem and make us stress free and immortal, for Godsakes.

Well, yeah, duh, of course.

She was revving up once more.

Okay, if duh, of course, why are we—both of us scientists!—kind of falling into that trap?

What do you mean?

Well, we seem to be acting like if we keep pushing for more and more information, we are going to get some certainty in the outcome. We aren't.

I started to say something snarky again but then thought better of it and held my tongue. She went on.

I mean, a hundred years from now some other couple is going to be sitting around ripped apart by what to do for their sick kid. And they are going to have options we can't even imagine right now. But their kid is still gonna be sick.

Okay... but what does that mean for us now? Not sure I'm following you.

I just mean that we, as much as anyone, should know that science has its limits. I mean, hell, it wouldn't be fun if it didn't. And yes, it's hard when those limits are about your own kid. But we aren't doing ourselves or Neva any favors by obsessing about this around the clock.

Bit by bit, over the years that followed, I began to understand more of what I think she was trying to convey in those moments and how her own seemingly boundless joy from science was partly because she embraced its limitations. Diana knew better than most that science at its best is what we should hope for in society as well but instead see fading away: the acceptance of flaws and limitations while also seeking to lessen them and the deeply held belief that something can be truly wonderful while also being far from perfect.

Science becomes its whole self, in both discovery and within us, when we don't try to deify it or demand perfection to give it our trust. When instead we view it as akin to the way most religions frame us humans: basically messy, hopefully striving every day to improve... and still capable of astonishing miracles.

CHAPTER NINE

Soon after Diana returned from Germany, we had another big decision to make, one that had begun a few months before. As our daughter slept in her final days in the ICU, Diana and I began to talk. Were we in the best place for her? What might she need? Would we have enough money to get her the best medical care?

At the height of these conversations, I got an email about jobs nearly two thousand miles away. Duke University was looking for a new dean of their School of the Environment as well as a new professor of microbiology. One was a good fit for me, the other for Diana. The jobs, should they work out, would mean a salary increase and close access to one of the best brain tumor centers in the world. Little did we know how much that would come to matter.

We agonized over the possibility of leaving Colorado, arguing with each other and ourselves. I took out my stress on another surgical resident, this one unfailingly nice. Later I felt like shit about it and bought him a fancy coffee and a double scoop of ice cream from the café downstairs. When I tracked him down outside another patient's room, he looked at me with tired eyes.

I'm lactose intolerant, he said kindly.

I mumbled something incoherent as he walked away; then I felt even worse. But I drank the coffee and ate the ice cream anyway.

That evening, we had sat upon the windowsill of Neva's hospital room, watching yet another extraordinary sunset. We could see the fading outlines of the Flatirons, slabs of ancient sandstone that protruded from the hills above our Boulder home and below the jagged outline of the Continental Divide. The sky contained tendrils of clouds whose underbellies turned from a brilliant orange to soft rose; one appeared spun from the northern ridge of a peak that sat unobtrusively beside its showier neighbors. That peak was our daughter's namesake. The view recalled the first night of her life, when we sat and held her upon another windowsill in a hospital just west of here. The night after I got the email, Neva slept with a clear tube snaking from the small of her back to an IV pole behind the bed, while we talked about this place we loved, about how it brought us together, about the community it held, and about the heartbreak it would cause us to leave.

Then we decided to apply.

On the heels of Germany, it became real. We'd been offered the jobs at Duke. We struggled with the decision once more, and then Diana returned to a point she'd raised about Neva's treatment options.

We can analyze this to death, but we're not going to know what's right.

I know.

We looked at each other for a few seconds, and then she took both of my hands in hers. I had a sense of what was coming.

I think we should go. We already decided it was the best thing for Neva.

Soon it was done. The moving van gone, the final hugs complete. On a crisp June day when the still-snow-capped peaks stood out in sharp relief against a cloudless sky, we began the drive east to a new home none of us knew and none of us really wanted. I felt pangs of loss and uncertainty as I strove for one last look at the receding peaks in the rearview mirror.

It was the second time we'd made the drive to the East Coast as a family. The first had come two years before, when Diana and I were both asked to take temporary positions at the National Science Foundation in Washington, DC. As we crossed out of Colorado this time, Neva asked, *Dada? Is Duke in Washington?*

No, love. But it's not that far away.

Does it have a Metro?

Neva had become a Metro addict during our time in DC, an offshoot of Diana's habit—begun when our daughter was only weeks old—of loading Neva into a jogging stroller and heading out for a run. There is an underpass in Boulder where a mural contains an image of a woman pushing a jogging stroller; it is of Diana, who passed the artist day after day. In DC, she ran from our rental house in Alexandria along the western banks of the Potomac, or at times across the bridge and to the National Mall, where I'd meet them for lunch or dinner before we headed home together on the Metro. We quickly learned to keep a close watch on Neva, because she'd try to pull away and jump onto any train she could, calling impatiently at us to follow.

Diana's role at NSF was leading the agency's program in Antarctic biology. It meant a chance to shape the science done in one of the most unique and important corners of the planet and a trip across the frozen continent that few on earth would ever take. But to go on that trip, she had to pass a physical, and that's when the threat of cancer hit our family for the first time.

Although she was not yet forty, a mammogram was part of the battery of tests required for her to receive medical clearance. More stringent mammogram guidelines were added to the protocol after a doctor who overwintered at the South Pole diagnosed her own breast cancer via a self-administered biopsy and then treated it with makeshift chemotherapy until the weather conditions allowed evacuation. Diana loved the story of the doctor's pragmatic bravery but bristled at what she viewed as yet another bureaucratic overreach. Especially one in which an agency whose mission was to advance science seemed to be ignoring, well, the science.

This is absurd. I'm not overwintering, for Chrissakes. I'll be there for three weeks, and there is absolutely no scientific or logistical basis for doing this.

She fought it, largely on principle, but ultimately conceded when the trip itself became imperiled. A few days later, I sat in a small Georgetown waiting room with tan couches and abstract paintings while she was taken back for the scan. I watched a gaunt woman across from me with a scarf upon her head struggle to hold up the book she was reading, and in a preview of days to come in our life, I felt a moment of fear.

The next day Diana walked into my office at NSF.

They found something on the mammogram.

I felt weak but managed to get out, *What did they find?*

Diana seemed unfazed and went straight into a scientific analysis of the situation.

It's small. Could easily be a false positive. Calcium deposit. This is what I meant when I said this test was bullshit.

Do you need a biopsy?

I don't know. She said to come in again next Monday, and they will do a higher-res scan. Take it from there.

She turned out to be right, but I couldn't help worrying as I paced the streets of Georgetown while she was in for the second scan. The usual masses of cars and cabs and people formed a noisy background watercolor, and I nearly walked into traffic at one busy intersection, my mind distracted by the ever-passing time without any news. Finally, a call startled me as I waited at another crosswalk. It was Diana.

False positive. Let's get the hell out of here.

I leaned against the steel lamppost on the corner while shutting my eyes.

Soon enough she was sending pictures from Christchurch in her official-issue red parka and giant boots and making funny faces at Neva over Skype from the bottom of the world. She told us both about spending a night in a snow cave she excavated and about helicopter trips to the Dry Valleys, where the tongues of giant glaciers gave way to a rock- and sand-strewn moonscape. She spoke of seeing Shackleton's Hut on Cape Royds and then of the seemingly endless interior of the continent as she flew to the South Pole Station.

A semicircle of flags from all countries in the Antarctic Treaty surrounds a ceremonial pole itself, the top of which is a mirror-bright globe. She sent a picture of her doing a side plank in full gear before the flags and another of her standing beside the globe with a grin I knew to mean she was laughing to the core. Later, I learned it was because she had leaned against the South Pole marker and nearly knocked it over. She came back from the trip energized and a little changed. Unsurprisingly, she immediately began to make plans to return.

Those plans came up again on our drive to North Carolina. In a typical fashion, I was driving and she had her laptop open. I was still stuck in the pain of leaving our home and worried about Neva, but Diana had found peace in her science once more. She brought me into the fold by asking, *Remember the Mexico project? I think I can do it better in Antarctica.*

She was referring to another trip the three of us had taken together, one of Neva's first plane rides at the age of only five months. We'd flown to Cancun, secured Neva in a battered Chrysler minivan, and pointed it south. We were bound for the tiny hamlet of Punta Allen, fifty-five rough kilometers past the bohemian tourist town of Tulum. The dirt road beyond Tulum bisects a long and narrow peninsula that forms the northeastern border of the Yucatan's Sian Ka'an Biosphere Reserve. The rest of the reserve can only be seen by shallow draw skiff or kayak, from which you can explore a vast mangrove-lined estuary, or on foot through the remote jungles to its west. It's all just beyond the garish hotels of the Riviera Maya but a million miles away.

We were there to trap bacteria in mason jars. *You have to go to Mexico to do that?* said, well, almost everyone. To simply fill a jar with bacteria, the answer was no. But to answer the question Diana was after, quite possibly yes.

She wanted to know something that once again seemed straightforward... but wasn't. Bacteria are almost everywhere—on you and inside you, filling those Costa Rican soils, even floating through the air in surprising numbers. But what happens when they *aren't* somewhere? In a place scrubbed of them for worries of disease or in a place newly formed in the world? How quickly do they get there? Does it matter who arrives first, or does it all sort out to the same end point no matter what? These were the questions Diana sought to answer, because doing so could help explain how the bacterial world worked, how to avoid unwanted ones, and how to tip the scales toward those we might want around.

She was starting in the Sian Ka'an because it was a fairly simple and consistent environment that would allow her to set up the experiment she had in mind. Put simply, that experiment was: put out the jars, open the lids, and see who shows up each day. At the end of it, how different would the jars be? And were there any rules behind those patterns? As in, would jars next to each other be more similar than those miles apart? Or would the whole thing be random? The area had steady winds off the ocean to the east that met a strip of similar forest, which ran for nearly the entire stretch of the narrow peninsula. We'd fill the mason jars with a sterilized soup of bacterial food, then place them just inside the seaward edge of that forest. Some right next to each other, some far apart. But every jar would sit in that same wind, under a similar set of vegetation.

In practice, it meant hour upon hour in the minivan, bouncing along a pothole-infested road. One of us would distract Neva while the other donned sterile gloves, carefully opened each jar, and then used a syringe to extract a small aliquot of the solution inside. We'd then empty the syringe into conical tubes that were sealed with orange caps and put the tubes in a cooler in the back of the van. When we'd get back to our rental cabin, everything would go in the freezer. From there, they'd get taken back to Colorado and analyzed in her lab.

She'd come to learn something fascinating from it all: that who showed up first was seemingly random but that those first bacteria to arrive had a significant influence on who stuck around later. Kind of like going to a party to find someone you love...or someone you can't stand. In the first instance, you're staying. In the second, you're getting the hell out of there.

The jar sampling consumed several hours of each day but left the afternoons to sit upon remote beaches of powdery sand or in the ocean itself while holding Neva in the gentle swells. Sometimes we posted up on the bridge above Boca Paila, an artery between a vast estuary and the open sea that separated the northern and southern halves of the Sian Ka'an road. A few years before, we had come here for the first time and parked our car in a pull-off just north of the bridge, then entered the shallow water on the estuary side. We had hugged the mangroves to our left as the depth increased to our chests, keeping an eye out for crocodiles, as the line of mangroves bent to the south and we could no longer see the road behind. We rose back to knee-deep water on the edge of a sparkling white sand flat, mangroves still to our left, a deep blue cut of water far to the right, acres of crystal-clear shallows in between. Not surprisingly, I had a fly rod in my hand.

Ghostly bonefish would come out of the deeper cuts and patrol the flat in search of tiny crabs, shrimp, or other meals scurrying along or buried in the sand. On that first trip, we paused near the edge of the flat, and Diana tried fly casting for the first time. Bonefish are not the recommended target for a novice fly fisher, as the winds are inevitably blowing against where you need to cast, and the fish are maddeningly spooky.

You have to be quick, precise, stealthy; even then you fail a lot more than you succeed. Those with years of experience chasing trout will sometimes scream and curse and throw a rod in frustration when they can't put the fly where it needs to be. Not knowing it was supposed to be hard, not knowing (or caring) how revered a bonefish was in fly-fishing circles, Diana soon relaxed into the rhythm of casting for the pure enjoyment of the motion.

Good God, I thought. *This might work.*

And it did. When a school of six came toward our chosen spot, she did everything right. A single back cast to load the rod, a forward cast punched into the headwind but kept just high enough for the line to land softly on the water's surface. The tan shrimp imitation settled to the sand below, and she began to retrieve it in slow, short pulls of line. One fish broke free of the school, pounced, and the line between rod tip and fly went tight. Then it began to scream off the reel while Diana held the bent rod with a look of pure and childlike astonishment.

We took Neva to the same flat on the last day of the jar sampling trip. I held her high as we stepped carefully through the soft mud beside the line of mangroves, the water lapping at my chest once again. When we reached the flat, I sat in the sand with Neva on my lap, letting her splash the water. Then I stood and balanced her upon my shoulders while I told my five-month-old daughter to look for her first bonefish. She was more interested in banging me on the head. Diana rolled her eyes.

Now we laughed at the happy memory before Diana unspooled her ideas for Antarctica and how they connected to what we'd done in Mexico. Her excitement was evident as she began to chatter at me about something called cryoconite holes.

What holes?

Cryoconite. They're these holes all over the surface of the glaciers that fill with meltwater. But they're so cool because they each form their own little unique ecosystem, totally surrounded by ice.

I was half with her, half still in memories of the Mexico trip, so I responded with a lame, *Ah yeah, cool.*

She steamrolled right past it.

Yeah, right? See, they could really test what I want to do. I mean, they are in the middle of nowhere, totally contained, and they repeat all across the glacier. Plus, you can figure out how old they are and maybe even make new ones.

Okay, tell me more.

She began to lay out detailed theories of how a community of bacteria might change over time, how much it might depend on who got there first, and how the simplicity of these icebound holes could allow more fundamental insights into the ways bacteria come together anywhere.

See, she said, *new holes form every year, and I could make artificial ones too. And I could even seed some communities—meaning intentionally start them with different combinations—while letting others just develop naturally. And do it year after year, so that I'd get a really good picture of bacterial colonization.*

Okay, I get it, but still, why Antarctica? I mean, I know it's awesome, but why there?

Because the communities will be super simple. Not much diversity compared to just about anywhere else. And that will help a ton in trying to figure out more general rules for how this all works. Most places have so much different stuff floating around that it gets tough to tease out.

She told me part of the work would be a lot like Mexico, but instead of jars and an old minivan, she and her team would need to access the sites by helicopter, then go from hole to hole taking the samples she required. Once again, she would track them across time and space, but in addition to just letting some of them develop naturally once they formed in the Antarctic summer, she'd also deliberately start a series of ice holes with bacterial communities of her design. Put it all together, and she hoped to unravel some underlying principles governing the assembly of these communities, ones that would perhaps translate to more human environments. The research she planned might give insight into everything from how to grow food with less environmental damage to how we might address a whole range of health maladies by shaping the bacterial communities in our own bodies.

Maybe even cancer, she said.

That one left us silent for a bit, but it didn't seem to dent Diana's focus. She went back to the laptop and soon enough was banging away at its keys. By the time we crossed into North Carolina, she had a first draft of an entire proposal and was bombarding me with ideas for how she would set up her new lab at Duke to make it all work.

Our destination finally came into view on a largely amorphous plain where the landmarks were road signs and the occasional city break. We made a brief run down a new freeway as a Gothic tower rose above a patchwork of green. We took the Duke Street exit, passed by the renovated shells of old tobacco warehouses, their interiors now bars, restaurants, apartments. You could sense hints of conflict and historical wrongs in the juxtaposition of hipster eateries and still-downtrodden surroundings. We drove a few more blocks, took a left onto Markham Avenue, then a quick right into the driveway of a Kelly-green corner house, its shutters a statement shade of purple. Our new home.

Within days, a pack of kids named Chambers raced through the house with Neva, and it became a pattern. The family lived across the street, and the kids would careen through one house or another until somebody yelled at them to go outside; from there, the destination was often a large and decomposing trampoline in the Chamberses' backyard. Its braces bore deep coatings of rust, and one of them bent sharply in a way that did not inspire confidence, while the black surface itself was encircled by ancient springs and leg-trapping gaps. Prior to Neva's tumor, the whole thing would have struck me as a welcome throwback to my own childhood. Now I had to fight waves of concern, even more so when the kids would amp up the danger by turning the hose on the tramp before jumping and slipping and falling in heaps amid screams of laughter.

Yet Neva remained unscathed. She had another good MRI, settled into her new school, and with each passing week, the trauma of her prior few months seemed to fade just a bit—for her and us. My new job took some adjustment, because I could no longer concentrate on my own science in the ways I loved. Now my primary role was to support the discoveries of others, of students and faculty alike. I tried to fit in with the new culture I'd entered but lasted only two weeks in uniformed navy blue or pin-striped gray before ditching it all for jeans and boots.

By the time the oppressive heat began to crack and the summer vanished in a whirlwind of student arrivals, I found myself tethered to a calendar as never before. Go here and talk to this group; run over there and talk to that one. I laughed in recognition when my new boss, a brilliant woman named Sally with a slightly mussed head of gray-brown hair, described her own job as Duke's provost as mostly consisting of showing up somewhere and saying a few words.

Meanwhile, Diana seemed to burst with energy, and she was constantly developing new ideas for testing the scientific mysteries she wanted to solve. One afternoon, I walked from my office to her lab and found her and Neva sitting beside each other on high black stools, each in a lab coat and safety glasses. Neva's coat sleeves were rolled up almost comically, and the bottom edge extended beyond her legs. My wife gave instructions as our daughter oh so seriously held a pipette in front of a row of the conical tubes with the orange caps.

As our first winter in North Carolina approached, I struggled with moments of jealousy and regret as I watched Diana refine her plans for Antarctica and develop new ones for Africa's Okavango Delta. My days in Costa Rica were probably done. I began to wonder if I'd made the right decision to step out of that world, even if it was for our daughter's sake. I loved being in the field, shotgun bruises and all. It was the place that lit up my own curiosity in ways well beyond just the questions I was trying to answer. At one point I told Diana I thought I'd made a mistake.

Hon, I don't know if this was a good idea. Me becoming a dean. I miss what I used to do. Feel like this is going to deaden my brain.

In typical fashion, she gave me a bemused and analytical look.

So go do it again.

I sighed and threw out my hands.

It's not that simple. There's no way I can keep working in Costa Rica while trying to be a dean.

Yeah, I know. I meant do this job for a bit and then go back to it.

But you know how it is. Once you step out of the research game, it's hard to get back in.

She looked at me silently for a moment, arching her eyebrows.

So that's a reason not to do it?

I smiled and acquiesced, telling her *you're right*, while inside still thinking I'd taken a one-way street and that an old life I loved was gone.

But then January came, and none of it really mattered anymore.

CHAPTER TEN

It started with pain in her right hand, during another trip back to Costa Rica. She was, as ever, so happily awash in her latest idea that even the occasional roar of a howler monkey in a guanacaste tree to her right and the sweep of the Pacific coastline far below barely made a dent. It was all just a watercolor background to the laptop screen. I sat across a cement patio and watched the familiar intensity of her gaze, intervals of stillness and furrowed brow punctuated by the staccato keyboard strokes. Her dark hair was still wet from a morning swim, falling across her cheek, turning the strap of her teal dress a shade darker as it dripped onto her shoulders.

For once I wasn't annoyed by her focus on work and instead was able to do something I loved—observe and wonder. I just sat and watched her, a slight smile on my face, as I asked myself: How does she focus like this? And why?

Only years later did I start to get the answers, or at least the ones I think are right. A good scientist knows that focusing effectively on one question requires ignoring others. Diana knew this, and so did I, of course; she was just better at putting it into practice. But I think there was something more. From the first moments she came to it, science was also Diana's escape. From a family in which she didn't quite fit. From a first marriage that was more convenience than love. And now from the horrifying specter of her only daughter carrying a tumor inside her brain. Diving into the details of a scientific problem wasn't just a source of intellectual satisfaction. It soothed her soul.

It took going through my own trauma therapy to learn some of the connections here. Distractions are a neurologically proven way to pull people from the depths. A good distraction such as a hobby or a video game can enable our courage and coping when we need it most. There are even seeming paradoxes in this truth: combat veterans who battle suicidal impulses often find relief from their PTSD in video games, including—sometimes especially—violent ones. Such distractions are literally saving lives. For Diana, I'm certain her escapes into science let her show up fully for Neva instead of being consumed by paralyzing fear. In time, I realized they also let her show up for me, and for herself, when the unthinkable came to pass.

Because on that bright tropical morning, something new was distracting her. I watched her shake her hand over and over and with exasperation, as though annoyed that a part of her would do anything but channel its energy into the science before her. Eventually I broke in.

Honey, what's wrong with your hand?

It hurts.

Her eyes remained on the screen.

Did you do something to it?

No. It just started hurting.

Since when?

She paused and looked up from the laptop with a brief sigh, realizing I was not going away.

About a week now, I think. It's been a bit worse since we got here.

We'd come to Costa Rica for vacation this time, to join a gathering of my family members for the holidays. We could hear Neva in the background, yelling at my dad to watch her jump in the pool. Diana continued.

It's okay. I've been writing a lot—I bet it's carpal tunnel or something like that.

So maybe you should take a break from writing?

What?

Her eyes were back on the laptop. I let her be and headed for the pool, trying to ignore a now familiar hollowness in my chest.

We were back in Durham a week later, in the midst of what had become a regular routine. This time we gathered at the Chamberses' house, drinks in hand, laughter flowing, and the din of the children oscillating between loud and deafening. Diana sat on a rotating silver barstool, with her elbows on the black counter of the kitchen island. The sink, refrigerator, and even the microwave were fire-engine red and evoked a bygone era of milkshakes and poodle skirts and paper hats. Diana listened to the stories and occasionally needled and flashed that gigantic smile. But she kept pulling her right elbow off the counter and shaking her hand.

Our friends noticed.

Hey, what's wrong with your hand?

Oh, I punched Alan earlier when he was being a pain in the ass.

Everyone laughed, and the evening continued as before, until we tore Neva away and put her to bed. As we sat beside each other in the living room, Diana confessed a deeper fear.

There's pain up to my elbow now.

What? How long has that been happening?

Last couple of days.

I took a deep breath.

Honey, you need to go see somebody.

I know. I called to make a neurologist appointment today. Hard to believe, but they can get me in next week.

Good.

She paused for a moment and looked at the long-sealed coal fireplace before us.

I'm a little afraid.

Another deep breath.

Of what? You don't think it's a repetitive motion thing? Those can suck, but I'm sure it can be fixed.

I don't know. I'm a little scared it's MS.

Oh God. Why?

I just have a feeling. There's some history in my family. And something doesn't feel right.

Now I, too, paused as I sought to breathe normally. Then I tried to reassure her.

Let's just get you in to see the neurologist. I bet it's fine. Like you said, you've been on the computer a ton. Makes the most sense it's from that.

Yeah.

She didn't look convinced.

When the day of her appointment came, I walked across the parking lot that separated my building from hers, up a long flight of wooden stairs, and into her office. Not surprisingly, she had lost track of time and was deep in Antarctic logistics and was annoyed about having to break away.

Can I have a few more minutes?

She shook her hand again as she said it.

I don't think so. We'll be late if we don't get going.

She sighed in resignation and closed her laptop but loaded it and a stack of journal articles into her backpack, and she couldn't resist another pushback as we walked out of her office.

You know we are just going to sit there. Probably better to work here. They won't even know.

She was still grumbling a bit when we walked into a single-story brick building a couple of miles down the road from Duke's main campus. But as we sat in the waiting room, she grew quiet. Then, *I'm so tired of hospital waiting rooms.*

Me too.

The neurologist was running late, to which Diana said, *I told you so*, but with a gentle smile. The delay forced me into a predicament that she settled for me. She knew the appointment might run right up against a scheduled meeting with my boss, Sally, and she insisted I go.

I have my laptop and need to work on this stuff anyway.

The provost's office was on the second floor of the Allen Building, one of many encased in the ornate Gothic sculpting of Duke's iconic gray stone. Sally shuffled through a stack of folders on a corner desk and extracted a notebook.

Damn it, some days I can't find anything! Does that happen to you?

I thought of my own office.

Uh, my desk makes yours look immaculate. So no, it never happens to me.

She laughed, and then we began to discuss the school's budget and our joint hopes that a donor's long-promised multimillion-dollar gift would finally come through. But as we dug into the specifics of how we might get out of a financial hole, I had a hard time keeping my focus. When she asked a question and I didn't even answer, she sat back.

Are you okay?

Not really. Diana's at a neurologist appointment right now, and it's kind of freaking me out.

Sally came to the provost job after years on the medical side of Duke's campus. She was a cancer biologist, one of the best in the world, and had also confronted plenty of her own medical hurdles. Her tone immediately switched to one of quiet concern.

Okay. Tell me more.

Well, she started having consistent pain in her right hand about three weeks ago. Then it began to extend up her arm even past her elbow at times. She also says that it feels a bit numb.

Was that a flash of worry in Sally's eyes? Or did I just imagine it?

Anything else? Or is it just her arm?

I think just the arm. Although she has a bad habit of hiding stuff until she can't.

Sally asked who the neurologist was, reassured me he was good, and then said, *Try not to worry until you know more. I know that's hard. Also, you guys really need to let me or Nancy know when you need specialty care here. We can get you past the standard gatekeepers, and we know who is best. You need the help, you call, okay? Anytime.*

Nancy was the dean of the medical school, and I knew they both meant it. They'd each reached out soon after our arrival to help with Neva's care. I nodded, grateful for her reminder and for her subsequent order:

Now, get the hell out of here and go be with Diana.

I arrived back at the neurology clinic just as she emerged.

He's not sure what's wrong. He said it could be something like carpal tunnel, but he seemed a little evasive on that. He wants me to get an MRI.

By now I hated that particular sequence of letters and blurted out, *Oh shit. Why?*

Well, he said it was kind of standard for these symptoms, that it could be a range of stuff. Including MS. But he also didn't seem super worried. He said there was no need to rush the MRI. They have it set up for March.

I felt a mix of fear and relief all at once. But within a week, it all turned to dread.

She walked into the kitchen, her face ashen and confused. It was early, and I was fumbling with the coffee maker while Neva still slept. I saw her, and the coffee was forgotten.

Honey, what's wrong?

I . . . I don't know. Something really weird just happened.

What?

She paused for a moment, as if trying to figure out how to answer.

I was in the bathroom upstairs, and I wanted to get a little piece of toilet paper to wipe off some of that medicine I put on this thing on my face—she pointed—*you know what I mean?*

Yeah and?

I . . . well, I don't know how to explain this. Basically, my head was telling me to reach down for the roll with my left hand like I always do. But my right hand is what reached out. I was just kind of pawing at the air with it under the sink.

Are you sure you're not just tired? Maybe you need coffee?

Knowing as I said it that I was in denial. Of something.

No. This was super fucked up. I'm telling you, I was standing there while my head was directing my left hand, yet I watched my right hand just wave around instead.

We held each other's eyes for a time, each in search of an anchor. Beseeching.

I managed to get out, *I think it's time to call Nancy.*

Diana just nodded.

When I got Nancy on the phone, she was equal parts action, sympathy, and comfort. But I could detect the worry in her voice too. Within an hour the March MRI was moved up. We were to show up at the hospital at six that evening and should know the results just after Neva's bedtime.

We skipped work and spent the day orbiting each other in uncertain paths, at times tightly bound, at others erratically drifting into different corners of the house. At one point I found myself before the upstairs bathroom mirror, gripping the sides of the sink and pleading with my reflection for her to be okay. With some unknown power to make it me if it has to be someone. I needed to pull it together but was struggling to do so, and I didn't know how I could leave the room.

But I did and found that Diana had somehow stabilized. She was at the laptop again, the science light in her eyes. She asked me a question about dust in the atmosphere and the nutrients it contains and then raised an eyebrow indicating she was not entirely satisfied with my halting answer. She probed for more details, explaining how dust might be a key source of essential elements to those Antarctic ice holes that were consuming her thoughts. I tried to steady myself and told her what I knew, including that ice cores drilled on the same continent might even give her a sense of how those dust inputs change with time. She looked at me quizzically after I talked about this, then asked a question.

Isn't that just over thousands of years? Like, it can't tell me what happened last year or the year before, right?

She was right. The record doesn't even start until years into the past, after the surface snow gets progressively buried and compressed into layers of ice. Meaning: for her experiment, my suggestion was probably useless. I took that in and tried to pivot to another idea, but I could not focus on the science, and my voice trailed off.

She stood up and spoke quietly.

It will be okay. No matter what, it will be okay.

That night Diana emerged from the MRI a little shaken.

God, I hated being inside that tube. I'm glad Neva is asleep for hers.

Did they tell you anything?

She looked at me with slightly widened eyes.

No. We have to wait to hear from the radiologist, but I asked the MRI tech if he could tell me anything. And he seemed odd about it.

What do you mean?

Well, he was totally avoiding my eyes.

Probably nothing, I said, as I screamed inside for that to be true.

We stood talking several feet away from where Neva sat cross-legged in a wood-framed chair with a floral print. She called out to us.

Mama, why did you have to have an MRI?

I'd already lost count of how many times she asked the question.

Oh, love, they are just trying to figure out why my hand hurts.

Giving the same stock answer that clearly did not soothe our daughter's anxiety.

Did they take pictures of your head too?

Yes, but only because our heads control our hands.

That seemed to help, switching Neva into a moment of discovery.

They do?

Sure, hon. Our brains control everything!

Neva's face tightened again.

Mama, is something wrong with your brain?

I don't think so, love. The doctors do this to be sure, but I think something is just wrong with my arm. It will be okay.

It wasn't. Hours later the little girl with the already-known brain tumor sat looking small and frail, shielded by a pair of oversize pink headphones, her eyes on the tablet's movie but her jaw noticeably tight. This was not a normal Friday night. Mom didn't get MRIs—she did. And yet there she sat, hopefully unable to hear, as the radiologist tried to hold Diana's gaze and struggled to say the words:

There is a large tumor in the left parietal lobe and another one down near the brain stem.

Diana had a pair of glioblastomas, a brain tumor that is incurable in all but a few miracle cases.

All I could think was:

What the fuck. How?

This woman who burst with life, who constantly brought it out of others, was now almost certainly facing death within five years, maybe far sooner. Half of glioblastoma patients didn't even make it a year. I didn't know if I could take it.

Yet as we walked out of the hospital, our daughter's hand firmly gripped in hers, Diana paused and looked at me with remarkable calm. With a small smile, she said:

This shit better not keep me from going to Antarctica next year.

That night was a blur, and the next morning was almost worse than the MRI itself. We sent Neva on a playdate and sat beside each other in bed, making call after call.

We have some tough news.

Some of our loved ones struggled mightily for words, leaving us all in painful silence, and at times Diana seemed to feel the full weight of the news for the first time. Yet others were instantly there in just the right ways, somehow able to table their own shock to be what we needed. We reached Greg and his wife,

Robin, as they touched down on a flight to Hawaii. Within hours, they boarded one for North Carolina.

And then one friend—maybe unintentionally, maybe not—blew the dread of the morning apart.

It began much like the other calls, Diana this time delivering the news. The line went silent as we looked at each other with a mixture of confusion and pain. Then the friend spoke in a rush of words.

Yeah, wow, okay, you guys, um, sorry, but am kinda scattered. See, I'm like literally standing here covered in shit.

We looked at each other again, this time only confused but a smile spreading across Diana's face.

Uhhh . . . what?

Yeah, well, see, I just got up and, um, sorry, not meaning to be gross but used the bathroom and the toilet wouldn't flush and so I got out the plunger thing and pushed down a bunch of times, and yeah, I pulled it out and looked in, and the toilet pretty much exploded in my face.

We lost it before Diana was able to ask the obvious question:

Why the hell did you answer the phone?!

The absurdity of the call seemed to return Diana to her typical self. She got up and announced she was going for a run. When she returned, she took out her laptop once more and soon enough was back into questions about dust in the ice. She rattled off half a dozen or more, each one building on the last, and her face brightened as she began to figure out how this particular piece of her research plan might fit together.

The Talmud teaches that our creative and curious selves are the God within, the slivers of eternal that reach beyond our mortal beings. As I was beginning to see, Diana's relentless devotion to scientific curiosity opened up that celestial part of herself when she needed it most.

It was present when the neurosurgeon sat across from us, surrounded by haphazard stacks of academic journals and clinical reports. A football signed by a famous NFL quarterback lay on the windowsill of his office, half-buried beneath a cascade of other paraphernalia, while the monitor behind him displayed the stark images of Diana's MRI. The intricate folds of her brain were replaced at the center of one image by an ill-defined vesicular blot, as though someone flung a bit of partially mixed fluorescent paint at the screen. The neurosurgeon, Allan Friedman, had thin and graying hair swept neatly across his skull, a square face, and blocky 1980s-era glasses to match. Incongruously, I thought of a *Saturday Night Live* skit about Chicago Bears fans. He looked at us silently for a few seconds, then spoke.

I can't cure you. All I can do is try to make your life better for whatever is to come. You need to decide whether that's worth it. Surgery may well extend your life. It also has risks. Now, what can I answer for you?

I sat mute and a little astonished. The man was direct to the point of unfeelingly blunt, or so I thought, but a glance at Diana showed she liked him. She smiled and asked a few procedural questions and then said, *Well, what the hell. I didn't really have other plans this week anyway, so let's do it.*

That didn't make a dent in Friedman's facade, but the next exchange did. It began with him asking, *How do you want to do the anesthesia?*

What? I don't know how to answer that.

Meaning do you want to be asleep or awake?

I can be awake?! You gotta be fucking kidding me!

There was a gleam in her eyes. Friedman cracked and laughed out loud.

Yeah, we can do it that way. It's actually better.

Oh, that's just awesome.

The surgeon shook his head in wonder.

Okay, then. See you Saturday. After that you'll be Henry's problem.

Henry was the other Friedman, the one we had spoken to first, just two hours after leaving the shaken radiologist. Word somehow got to Nancy and from there to the man who led Duke's renowned brain tumor center. He had called that first night as we lay in bed, clinging to each other and still mostly unable to speak. His torrent of words and manifest confidence were a stabilizing force after the MRI. Near the end Henry ran on:

Look, this sucks, and I'm not going to sugarcoat it, but you're young and, from what I hear, you're tough as hell and we are good at what we do and I do not give up and so don't you either.

He finally paused for breath.

I'm going to see you first thing Monday. You'll see Allan in his office Sunday. He looks like a real doctor. I don't.

He didn't. Faded jeans and battered running shoes. A white Duke hoodie with a stain on the front. Unkempt salt-and-pepper hair with a beard to match. Henry swept into the little clinic room on the third floor of Duke's cancer center and immediately took Diana into a bear hug. He did the same to me. Then he began to lay out the treatment path ahead.

Diana broke in, her inner scientist taking over once more.

Isn't that just the standard approach? The one that probably only buys a few months? What about immune therapies?

Henry paused. Then he smiled at her and laughed.

Scientists. You're all a pain in the ass. I was getting to it, but chemo and radiation is not our last shot. It's our first one. It'll slow things down while we figure out the specifics of your tumor and how else we can go after it.

As we left the cancer center that day, Diana gave me a look of determination I knew well.

I'm going to see Neva finish grade school. And then middle school. And then high school. I'm going to see her graduate from college.

CHAPTER ELEVEN

It might not have helped with her experimental plans, but beneath the surface of the tiny glacial holes Diana hoped to study, there's a story of the world written in the ice. Each year, stretching back long before humans came into existence, a new frozen layer was formed. As the ice hardens, it traps a record of its moment in time to be held in near perpetuity. What was the atmosphere like back then? How much dust was flowing off the world's deserts? How high or low were the oceans? How warm or cold was the planet? Today we can pull a core of ice that is a history book of Earth's past environments, dating back nearly one million years. I love how Robert Macfarlane describes it in *Underland*:

> Ice remembers forest fires and rising seas. It remembers how many days of sunshine fell upon it in a summer 50,000 years ago. It remembers the smelting boom of the

Romans, and it remembers the lethal quantities of lead that were present in petrol in the decades after the Second World War. It remembers and it tells—tells us we live on a fickle planet. Ice has a memory and the color of this memory is blue.

I saw one of these ice cores for the first time in a bitterly cold room tucked into the back of a nondescript government building in Golden, Colorado. I wore a red puffy parka and thick white gloves while a glaciologist friend extracted one of the cores from the frost-bound storage racks, then showed me the banding of dust along its burnished cylinder, the simple pattern belying how much information on our world's ebbs and flows it contained. I wanted to reach out and touch the ice but hesitated, feeling it was akin to some off-limits work of art.

Later, I watched another man take a portion of the core and use a modified table saw to slice the cylinder into little discs that looked like semi-translucent coasters. Eventually, he'd melt parts of those coasters in a sealed container and measure the carbon dioxide and methane concentrations of the tiny air bubbles within the ice. Others would take a little of the dust in each of the discs and measure its radioactive uranium content. Do that, and you know how old it is. Combine it all with measurements of the isotopes of hydrogen and oxygen in the ice, and you can reconstruct the average temperature.

Assemble all those data and you'll find there are eras of relative stability, punctuated by moments where everything changes.

A few thousand years later, the whole thing reverses, sometimes with astonishing geological speed. One of the most remarkable such records is from a period known as the Younger Dryas. After several thousand years of warming as the world emerged from the last major ice age, things rapidly reversed course and parts of Earth suddenly turned much colder and drier. This is best seen not in Antarctic ice cores, but in similar ones from Greenland, because northern Europe took the biggest hits. Average temperatures in places like Great Britain, Scandinavia, and France took a nose dive, with disastrous consequences for human agriculture of the time. And it all went down in a hurry: the data suggest that the temperature drop occurred in less than a human lifetime, stuck around for more than a thousand years, and then things warmed back up even faster. It likely happened in the first place because massive volumes of meltwater from northern ice sheets shut off a major source of heat to Europe that comes from North Atlantic currents. When the melting stopped, presto, the warm currents came back.

These abrupt climate changes, visible in the ice, are one of the things that worry climate scientists the most, because the very latest records show that humanity's imprint dwarfs all past upheavals. It's bigger, it's faster, it's like nothing else we've seen for a million years. We are racing headlong into a future that has no prior analog. I'd taught classes about this for years; now I felt like I was facing it in my own life.

Scientists seek patterns and predictability. We do that via repeated observations, coupled with experiments designed to reveal what drives the patterns we see. For example, scientists who specialize in meteorology have become remarkably good at predicting where a hurricane will go and how strong it will be. They've done this by watching the patterns of past storms and teasing out what made them behave as they did. What was the ocean temperature? What were the pressure gradients in the atmosphere? And so on.

I can't lie. When our scientific predictions are right, it feels good. But we're not always right, and there are some things that are extremely difficult to predict even with mountains of data. This is a paradox of scientific endeavor—we must embrace life's fundamental uncertainty while simultaneously living in a body and world that so wants certitude. And sometimes we can't help but hope for something that defies all the data that allow us to make predictions in the first place.

That's where my mind was as I sat on a window ledge in the kitchen the morning of Diana's surgery, staring blankly at our back patio's koi pond in the soft predawn light. I couldn't stop myself from asking a question that no longer had any relevance. What was the probability of my wife and daughter both having brain tumors?

Neva's alone was exceptionally rare: of the roughly seventy-five million kids in the United States, only three hundred per year are diagnosed with a craniopharyngioma. Glioblastoma is—tragically—far more common, but much less so in women of Diana's age. For her, the odds were about one in one hundred and fifty thousand. But both tumors in a mother-daughter pair in just over a year? Best I could figure it, those odds seemed to pencil out at less than three in one hundred billion. That staggeringly low probability led to another calculation: chances were, Diana and Neva just became the only members of their fateful club in human history.

As I confronted these impossible odds, my atheism wavered. Now I felt like maybe God existed, and he was an asshole. How could this simply be bad luck? How could this not be some kind of karmic event or operatic tragedy? I couldn't brush this fatalism away; nor could I avoid a growing pull to believe in a higher power. In someone or something that could reach down and swat aside the devastating odds against my wife and me growing old together.

I was, in short, having a crisis of faith, one ironically similar to those that religious types often face during catastrophic struggles. Their lives crumble as they come to think the truth they worship must not exist. I came to a similar conclusion about mine.

Part of thinking like a scientist involves embracing the realities of chance and probability, which are embedded in any predictive framework, and ideally, finding some peace amid their vagaries. Much of our daily lives, from the ordinary to our most defining moments, is shaped by probability and thus can, in a way, be quantified. What are the odds that this pregnancy will occur? That the baby will be born healthy? What are the chances that the stock market will go up, that this player will make a crucial free throw, that it will rain tomorrow, that your car will run trouble free? Sometimes we have pretty clear odds we can attach to life's questions, sometimes not. Either way, degrees of uncertainty govern our entire lives. But we are always trying to reduce those unknowns.

Take chaos theory. At its heart, this is a mash-up of physics and math that illustrates how many of the complex systems that define our world can be sensitive to initial conditions and that there are some remarkable degrees of self-organization and pattern in things that we once thought were random. In some ways, Diana's projects in Mexico and Antarctica were exploring some broadly similar concepts that bleed over into ecology: namely, was the eventual bacterial community you might find in a given place highly sensitive to *its* initial conditions?

But chaos theory has an appeal that at times has dragged it far beyond what it actually says. In some public translations, it's become a way to connect every dot from the beat of a butterfly wing to the tornado that just ripped through your town. This, by the way, annoyed the shit out of Edward Lorenz, the so-called father of the whole thing.

Why did Lorenz's ideas go prime time and so far out of their lane? Because we *so want* that predictability in our daily lives. We want to know what a thing that happens over there means for me right here. So, we cling to numbers and odds and erroneous extensions of science into realms they never could—and likely never will—predict. It's frequently just an outfit to mask the underlying truth: we can only know and do so much. Sooner or later, something is going to happen that we can't control and can't predict, and if we spend our lives clinging to the odds, it robs us of the daily magic of our existence.

Still, I admit it's damn hard to find much magic when the person you love is facing a probable death sentence.

A house full of friends and family had descended upon us, and as they awoke, I left off staring at the koi pond and helped Diana pack a small blue duffel. Our houseguests tried to envelop Neva in a constant stream of distractions. They would work for a few minutes, and then Neva would ask yet again.

Why is Mama having surgery?

Eventually she broke free to go find Diana.

Mama, are you going to be okay?

Somehow, Diana found the equilibrium to sweep our daughter into her arms and swing her around in a complete circle before tossing her onto our bed. When that failed to strip the worry from Neva's face, she went for the ticklish spot beneath her chin until a laugh finally broke through. Only then did she answer the question.

Honey, the doctors are going to help me just like they helped you. And I won't have to stay in the hospital as long as you did. You get to have a couple of sleepovers with the Chamberses, and then I'll be home.

Neva made a face.

No fair! I had to stay two weeks!

It was a brief moment of lightness, but soon it was time. The gown, the repetitive vitals, the seemingly endless moments as we waited for the surgery itself were too familiar, too laden for us to do much more than hold hands and look at each other. A physician assistant with bulging arms and a surprisingly soft voice entered the room, shaved the left side of Diana's head, and arranged her beneath a laser array. The laser's red dots guided his inking of a series of Sharpied blue *X*'s on her temple and newly shorn scalp. He explained that these were part of setting up the surgical process for the precision it required, but the image of her shaved head with its blue marks hit me hard.

Then came a déjà vu moment from Neva's surgery, when the seemingly interminable wait flipped to a moving gurney and severed handholds and she was gone. I walked slowly from preop to yet another waiting room, where I recalled a children's book Neva loved. It's called *The Mitten*, and the gist of the story is that a boy drops a mitten in the snow, which in turn becomes a refuge for a physically impossible collection of animals. A tiny shrew and a mouse. Then a bunny, an owl, a fox, a wolf, a bear. Somehow the mitten holds, falling apart only when an achy old cricket seeks to wedge into a final gap. I sat, and the images of this book juxtaposed with those of Diana's brain, two shockingly large tumors somehow shoving their way in on the sly. Henry had explained that the brain has astounding plasticity, able to bend without breaking far more than one would think. Up to a point. So I pictured the cricket.

And I paced. Along the glass-wrapped external hallway of the hospital's sixth floor, then out the doors below and beyond the Gothic walls of the medical school, until I descended into the winding paths of Duke Gardens. The place is beautiful and peaceful, but like many such gardens, it is a diorama of examples from around the world that are nothing like the real ecosystems from which they come. As I walked, the artificiality of my surroundings outweighed their beauty. They reflected my own world, which I felt was now fragile and bordered. That thought chased me back toward the hospital waiting area.

When the time finally came, Greg accompanied me into a windowless room where Allan Friedman already sat at a small conference table. He was as direct and unceremonious as ever.

She did fine. No problems. You know, of course, that we could only take the parietal lesion.

Her second tumor, the one encircling a portion of her brain stem, was beyond an acceptable level of surgical risk. Even for him. Even for this.

I know.

You can see her soon. Any questions?

I struggled to come up with one, caught short by the surgeon seeming to conclude the meeting even as I was still finding my seat. But I managed the obvious.

Did you get all of the parietal one?

He softened a little.

You never get everything with one of these. Their edges radiate in ways that are nearly impossible to see or remove. That's what the chemo and radiation try to control. But yes, we got as clean and complete a removal as we could hope for.

Thank you. It was all I could say.

Allan looked at me through his blocky glasses for several seconds before breaking character even further.

She's funny. And tough. Both are good.

Funny?

She was making jokes in there. She also wanted to know how everything worked.

Then he shook his head with a little smile.

Check at the desk, but you can probably go back and see her now.

She lay quietly at a slight angle while her vitals traced and beeped on the monitor above. An IV line ran from her left hand to a softly humming pump and then to a nearly deflated bag of clear fluid above. There was a wrapping on her other forearm that defied easy explanation, but I spent little time pondering it as my eyes were drawn to the much larger bandage above. It encircled her head and covered the left ear but not the right, like some mummified flapper hat. A shock of tangled dark hair was still visible behind. Her eyes were closed.

I pulled a chair to the bedside and watched her eyelids periodically twitch beneath a furrowed brow. Subtle eddies of a deeper disturbance. I could not help but reach out and rest a hand lightly upon hers. She woke and looked silently at me for a long time without expression until I began to panic that she could not see. Then came a slight smile.

Does my hat look good?

I managed to say, *White is not your color.*

She drifted off once more as my already thin veneer of composure eroded beneath waves of self-doubt.

Could I do this?

For her, for Neva, for me. Could I somehow prepare for the near inevitability of a life without her, for the unspeakable devastation of watching her die, while simultaneously pouring every possible ounce of hope and belief into a narrative where her singularity extended even to this? I didn't know.

Another chilling image of the cricket bursting open the mitten forced its way into my head. It triggered a sudden fury, and I left in a rush, stopping only when I was in the locked confines of the nearest patient bathroom. I cursed loudly and punched the wall beside the mirror, almost reveling in the shooting pain that erupted in my right hand and forearm while I considered ripping the crimson alarm cord out of the fucking antiseptic wall. Then I stopped short at a knock on the door.

Sir? Are you okay?

The nurse was young with wide eyes and a freckled face. She took a step back and looked at me warily when I opened the door.

Are you all right, sir? Do you need anything?

With that my anger melted into shame.

No, I told her. *I'm so sorry. Just a shitty week.*

Unmoored, I drifted out of the ward and ended up sitting on the upholstered arm of a waiting room chair, staring vacantly at the quad below. A gaunt man in a hospital gown had pushed his wheelchair onto the grass beside the path that led from the cancer center to the parking garage. His chin rested in his left hand while a tendril of smoke rose from the cigarette in his right. A few minutes passed, and then a young woman in jeans and a fitted tan pea coat trotted from the cancer center doors, a faded gray ski jacket under her arm. She thrust it back and forth before the wheelchaired man, her head bobbing sharply. He didn't move. She stood and stared down at him for a while, then draped the coat around his shoulders, carefully tucking it between his body and the arms of the chair, before taking the cigarette and flattening it beneath a brown leather boot. When he still did not respond, she slowly retraced her steps before stopping to look back. Her shoulders were slumped. So were mine.

But when I returned to the recovery room, Diana was awake and deep into the theories behind her Antarctic project. Her bright eyes were pinned on Greg, the somewhat astonished object of her attention.

See, there are these rare bacteria—don't know what they do; they never hit big numbers. But if this one colonizes the spot right out of the gate, then you get one kind of community; get a different colonizer, and you get a different community. And that can matter so much!

Greg shot me a *can you believe this shit?* look before trying to engage.

Okay, so is it random? Meaning who gets there first?

Her answer was quick, her expression lit.

I don't know, seems like it must be. But how cool is that, right? A random draw of rare occurrences sets the course for almost everything that comes next!

She failed to see the analogy to her current surroundings, or perhaps didn't care. But I did, so I sat quietly and watched the two of them dive ever deeper into the science of it all, making connections between Diana's bacteria and Greg's tropical trees, teasing out commonalities and even planning a paper to write. All of it just hours after her head had been split open and a portion of the insides removed.

And I thought, what are the odds of someone reacting like *this* to an incurable brain tumor? They have to be pretty damn low, right? So maybe she was the outlier I so desperately hoped for. Maybe her brain was just too consumed with scientific wonder for anything to stand in its way.

CHAPTER TWELVE

Near the end of the trail Diana and I first ran together in Costa Rica, there is a tree that has no business being where it is. It clings stubbornly to an outcrop of lava that gets clobbered by the sea. Wave after wave smashing the rocks and coating the little tree in salt. You might be picturing it as some noble banzai against this wild background, but in truth it's a homely little bastard that still manages to claw out a living against the odds. It's gnarled and misshapen and vaguely reminiscent of Gollum in *Lord of the Rings*. Its ugliness is all the more striking because of the breathtaking grandeur of the forests nearby. But I love the damn thing. I think about it often. It's funny and unexpected and inspiring all at once, and it reminds me that we are capable of finding footholds in life that we don't believe are possible.

My field of science can help explain how that tree manages to pull it off. Its roots are stronger than you would think, and they find ways to both anchor the tree in the lava and mine the nutritional elements it needs from the rock. In all likelihood, the tree also gets some of what it requires from the sea salt that plagues it: calcium, magnesium, potassium. And it deals with that salt by being a hardened little beast, including right up to its leaves, which are encased in shells of protective, waxy compounds that serve as little salt-spray umbrellas, while still managing to collect energy from the sun. Walk into the neighboring forest just a bit from this tree, and you'll find others of the same species, but they look nothing like it. They have it easier up there, so they can rest in showy luxury. And they are beautiful to behold. But I love their twisted cousin on the coast best.

The tree popped into my mind when I drove Diana home from the hospital and watched the deliberate placement of her feet on each of our front steps. They were stained dark from years of benign neglect, reminiscent of that lava point, and now they lay encased in a coating of ice, so I held her arm as she climbed slowly toward the covered porch and kitchen door. The porch eaves were adorned in purple, gold, and green Mardi Gras decorations, and the doorway was blocked by a welcoming committee with our daughter at the front. Neva leaned back nervously as she eyed her mother's bandaged head, but that soon gave in to a smile that radiated happiness. They came together as everyone, Diana included, tried to mask their wet eyes.

We all swept inside in a rush, escaping the unseasonable cold, bumping against each other through the narrow kitchen passageway before spreading out into the living room with everyone talking at once.

How are you feeling?

Mama, why do you have that white tape on your head?

Is anyone hungry?

Holy shit, I thought North Carolina was supposed to be warm!

Hey, give her some space!

The rest of the day was a blur of nervous energy and feigned normalcy until finally it was just she and I once again sitting beside Neva in her bed, reading her a story, soothing her to sleep. The routine was prolonged by all of us that night.

Only when Neva was asleep did Diana let her exhaustion show. She leaned heavily on me for the short walk to our bed and struggled to keep her eyes open as I brought her a trio of pills and a glass of water. Then they opened fully.

I'm going to beat this.

I could only nod.

The next morning the temperature was stuck in single digits, and yet she insisted on a walk. Friends and family members urged her to reconsider.

What if you slip on the ice?

Don't you want to rest just a bit more?

She ignored them and put on her boots. I did the same, and when we reached the end of the agreed-upon block after walking slowly in silence, she turned to me.

One more.

She conceded that I could hold her arm. We began to climb Markham Street, and she faltered momentarily on the hill.

Honey, that's probably enough for today. Why don't we walk back?

Sure. Go ahead. I'll meet you back there.

Her breath hung in the frozen air, and I recalled a moment a few years before, when I had picked her up at the end of a long run on a windy, subzero Montana day. Her eyelids were completely white with frost.

She didn't miss a day of walks. By a week out, she wanted to do them alone; by the second week she was running again. On the first of those runs, I stood watching her jog slowly up Markham, then slipped into the car and tailed her from a block away. When she reached Buchanan, the street that divided our neighborhood from the slightly decaying gray rock wall, red-brick buildings, and sweeping lawns of Duke's East Campus, she turned left and out of sight. I figured she was headed for the nearest gap in the wall and from there to the graveled running path that encircled the campus grounds. I waited a few seconds to let her cross the street before pulling up to the intersection, only to find her standing just down Buchanan with her hands on her hips and an accusing glare beneath a head no longer wrapped in gauze. The jagged scar that ran from her temple to behind her left ear was proudly on display. Sheepishly, I rolled down the window so I could hear her question.

What the hell are you doing?

I just wanted to make sure you were okay.

She softened and said, *Why don't you run with me?*

Then she couldn't help but add, *If you can keep up.*

Twice each week we returned to the cancer center, visits that sometimes included hugs and verbal torrents from Henry.

Okay good God you are out running already unbelievable now here is what we are doing next. Genetic testing of the tumor, get some more info, will help me think about what other therapies to try, got the idea of what I think we should do but want to check these other options, also this is Minh, you can tell her anything you would tell me and trust her with everything she's as good as it gets.

A young Vietnamese woman by Henry's side smiled in knowing tolerance.

Nice to meet you guys, is all she could get out before Henry continued.

Right, now, Minh will see you every week when you check in and I won't be here every week just can't be but am always in the loop and you have my cell phone and I don't give that to everyone and you can call it any time and I damn well mean it plus like I said she knows what the hell she is doing.

Minh smiled again and shrugged a little at Diana, who laughed in return. Then Diana turned to Henry.

Are you still thinking of the dendritic cell vaccine?

She meant an experimental treatment in which cells involved in the brain's immune response were spun from a patient's own blood and then turned into a series of vaccines in which the dendritic cells were hopefully boosted and tuned to attack glioblastoma cells. Like nearly anything with this particular cancer, the early results were no miracle—yet still hopeful.

Yes, that's the plan, but I want to do the genome work first just in case.

Two more hugs and he was gone. Minh remained to go through the logistics of the next few weeks, and when she finished, Diana asked about the seizure medication she was put on after surgery.

What about the Keppra? Can I get off that soon?

Have you had any seizures or anything that just feels like it could be seizure activity?

No.

Okay, good. That decision needs to run through neurosurgery, but I'll check. When you hit three weeks out, you should be able to go off it.

We thanked her as we all walked out the door of the exam room. Diana and I continued out the doors, then across the path where I had watched the wheelchaired man smoking. As ever, reminders of the building's purpose were there. Bandanna-covered heads, gaunt frames, pink T-shirts, and soft voices. We walked slowly to the parking garage, my right hand in her left. She shook the right one frequently, because the pain that had triggered her diagnosis in the first place was still present.

A week later and the same walk had an element of minor victory. Diana turned to me with a smile and said, *I can drive again!*

Allan had given the green light to get off the seizure meds. In her face I could see the lifeline that represented, the symbolic importance of another piece of her former life now back in place. She followed up her proclamation with another.

And I'm driving home.

I pushed back.

You're not actually off the Keppra yet, and you're supposed to make sure nothing happens when you do go off it before you drive.

By now you'd think I would have known better. The familiar set to her jaw returned as she paused before the parking garage elevator.

Tough shit. It's a mile. And I'm doing it.

Then a challenging smile as she said, *You can ride along or walk home.*

I raised my hands in surrender and gave her the key. At first the manual transmission seemed to stump her stricken right hand, but soon she settled in and began to push the car on our short trip home. A corner just a little too fast, a downshift not really needed. A look of peace on her face throughout.

For the next three days, she violated protocol and took any excuse to get behind the wheel. We took needless trips to the grocery store, to her office, to a local coffee shop. On the fourth day, she gave up the pretense and just said, *Can we go on a long drive?*

We left the city behind as she steered the car along a web of country roads. A new prescription had eased the pain in her right arm, and she guided the shift lever with confident ease. Music played as a patchwork landscape of loblolly pines, oaks, maples, and sweetgums, all interspersed with the occasional small farm, passed by. At one point we emerged from our tree-lined course to look down upon the twisting shoreline and gently rippling waters of Falls Lake. We scarcely spoke, each of us soaking in the unexpected gift of tranquility.

That night we lay beside each other, books in hand. The peace of the drive had remained and my mind relaxed into the novel, so I didn't notice that she had set her book down. Only when I heard her take a sharp breath did I look over to find that her right arm was above the covers and twitching as though electrified. She was staring at the ceiling as her breaths came faster and more labored.

I bolted up and grasped her shoulder.

Honey, what's wrong? Are you okay?

She did not answer as her entire body was now seemingly racked with the effort of breathing, desperate intakes as though the room no longer had oxygen. Her eyes were impossibly wide, and the breaths themselves became near screams as her body began to convulse. I found that I was yelling her name, and then, not knowing what else to do, I slipped behind her and simply held on as her body bucked against mine.

▲ ▲ ▲

The horrible rhythm of wave upon wave of convulsions seemed to last forever. I continued to hold on as I stared at the olive paint and white-framed window of the far bedroom wall, all of it becoming a blur. Then she grew quiet but noticeably stiff in my arms. I slid from behind and set her down gently. Her eyes remained open and wide but appeared unseeing. I braced my arms on either side of her, beseeching, yelling, staring into those eyes from inches away, then shaking shoulders that no longer felt animate.

Diana! Can you hear me! Diana! Are you okay? Are you okay? Are you there?

When she did not respond, I collapsed on top of her, disintegrating with the certainty that she was gone.

She was not. Her body shook briefly, as though experiencing an aftershock. Soon her breathing began to settle into a deep but normal pattern. I braced above her again and saw that her eyes were now focusing.

Honey? Can you hear me?

She didn't answer, but I could read the recognition in her frightened gaze.

Can you nod?! Can you nod if you hear me?

Her chin dropped slightly toward her chest; then she whispered *call Henry,* and I scrambled for the phone. It was just past eleven, but he answered on the third ring. By that time her voice was returning.

Put him on speaker.

Henry's typical rush of words was absent, as if he sensed my panic immediately. He was simple, direct, calming.

Tell me what is happening.

As I began to describe the events, he interrupted.

Diana? Can you hear me?

Yes. I'm okay.

Soft but clear.

Good. Look. You had a seizure. It's okay. It happens. Scary as hell. Does anything feel wrong now?

I don't think so. I don't really remember what happened.

Then he led me through a series of simple tests, pronouncing after they were complete that all sounded fine, that she should get back on the Keppra and that we should just try to go to sleep.

Come in tomorrow to get checked out. But I don't think you need to go anywhere tonight.

Henry, I thought I lost her.

I choked it out, and she grasped my arm.

I know. I know. These can look awful, but it's okay. Just don't go running in the morning, promise?

And with that he was back to office Henry, somehow weaving together a complicated monologue on smartwatches and dogs. I got the feeling he knew exactly what he was doing because, by the time he hung up, we were smiling once more. Diana returned to reading her book, *Endurance*, a story of Ernest Shackleton's Antarctic voyage. How appropriate, I thought. But she acted as though there was nothing really to endure, having flipped right back into fascination mode. She waved the book in my direction with a look of familiar excitement on her face.

Can you imagine going there back then? Damn, how cool would that be!

As she drifted off to sleep, I thought once more of the little tree.

CHAPTER THIRTEEN

Was Diana's response to her world being turned upside down just manufactured bravado? Perhaps, in part. But as the weeks passed, I realized it was more than just that. She was stabilizing herself by resting upon the familiar, by refusing to let go of habits that soothed her mind and kept her at peace. And by doing so, she was able to share some of that calm with us. So that what could be our final months together wouldn't be wasted in rage and despair. How did she do it? I think the neuroscience of stress and creativity gives some clues.

Our brains are packed with all kinds of different centers of activity, each one with a daunting list of tasks to perform and linked through a series of brain-wide networks. Sometimes those networks all seem to work together as a single, well-oiled machine. And sometimes it seems as though they are shouting obscenities at one another while every alarm in the place is screaming red.

In a recent review, psychologist Oshin Vartanian and colleagues describe one common conceptual model of the brain, in which we house both a "default network" and an "executive control network" in our heads, and when things are all fine and cool, the two work together to help our brains spit out creative insight: new ideas, concepts, breakthroughs. Data show that the default network cranks up more when our minds are happily drifting and at peace from varied external assaults. But once any shit hits the fan, the executives (as ever) tend to start throwing their weight around and reallocate the brain's activity toward specific and immediate tasks.

A third division, known as the "salience network," also comes into play. In my own mental model of it all, one that I'm sure would drive my neuroscience colleagues batty, I picture the salience gang as shadowy types who suddenly show up when you least expect them and who everyone secretly believes is actually in charge. In reality, this part of our brain both takes in external signals and triggers the allocation of mental resources to deal with them. Among other roles, the salience network appears to coordinate switching the relative dominance of the default and executive control arrays, and when the chips are really down, it sounds red alerts for our focus to be on survival... not on some daydreamed and undoubtedly half-baked new idea.

In most of us, the onset of stress does just that: shoves aside our creative and curious brains. We become Sapolsky's no-headed chicken. But there are exceptions—not everyone descends into fight-or-flight mode. When I think about Diana, I'm convinced that her daily habit of having a brain immersed in the wonders of science helps explain how she reacted to her life being on the line. Much like her daily runs built physical fitness and resilience, her brain was well tuned to dole out regular doses of the feel-good hormone dopamine in response to being curious and creative. So perhaps her default network was better at telling the executives to shove it, simply because it was so used to being in charge. It's not crazy or wishful thinking to believe this: brain imaging data show that the default network is unusually active in people who demonstrate clear and consistent creativity.

The seizure did deal a blow by stripping Diana of her brief interval of freedom, and for a couple of days she was not herself and the stress of everything finally appeared to take hold. Yet, not surprisingly, she soon brushed that aside, and we began to settle into a new rhythm of sorts. Mornings grew softer each day as North Carolina's early spring took hold. We'd sit and drink coffee on the wraparound porch, occasionally smiling at a Chambers kid yelling *HI!!!* at full volume from across the street. Early morning be damned.

Within days of the seizure, she was running again, slowly the first time and then with increasing sureness. Once more, she doubted my ability to keep up when I asked to go along. She returned bathed in sweat, showered, and then wrapped a brown and white bandanna over her partially shorn head. And as she did most every day, she crammed her laptop and a stack of journal reprints into her aqua-colored backpack with its Antarctica badge.

Are you ready yet?

Antsy to get to her office.

The pattern repeated nearly every morning. But so too did the daily walk across campus to the cancer center, where we would take the stairs down to the lowest level and then wait among ashen faces and forced smiles until she was ushered beyond the double doors to have another dose of radiation enter her brain. Our walks back were always slower and reflective. The whole thing never took more than an hour, yet the cycle would often affect the rest of my day. Unlike Diana, who seemed to switch right back into her science once she returned to her office, I'd remain stuck in thinking about our new reality, drifting in meetings, unable to focus.

By April it was better. The end of the daily radiation blasts was in sight, and she weathered the first round of chemo without too much discomfort. It even brought a moment of comedy. The chemo protocol consisted of short but high-intensity monthly blasts of the drug, and Minh recommended a THC-based medicine that eased the nausea for some patients.

Diana gave it a try on a night when I ended up in the ER for mysterious stomach pains that turned out to be diverticulitis. After a few hours of discomfort and nervous uncertainty given our recent track record, I left the hospital, gave my ticket to the valet, and then sat on a metal bench alongside the ER loop. My phone rang.

HoneyAreYouComingHomeSoon?!

Diana was breathless, too loud and rushed. I panicked.

Yes, right now—what's wrong?

It's that pot medicine Minh gave me!

I relaxed a bit.

What? What do you mean?

I don't know I'm just totally freaking out! Can you stay on the phone with me until you're home?

I tried not to laugh as I said *sure*. Later, we both lost it as she told me how the paranoia suddenly hit her in the Chamberses' kitchen. How she rushed wild eyed from the house, Neva in hand, and then had to get her young daughter to unlock the front door because she couldn't make the key work. How she somehow got her to bed and then huddled in a corner armchair, the dog in her lap, the shades drawn, debating if she should call me or not because who knows who might be monitoring her calls?

Throughout these ups and downs, she buzzed almost daily about ideas for her Antarctic project. And took her runs, which I later learned made her faint to the point of disorientation, but she waved this aside so thoroughly that no one knew. The truth only came out when she ran a benefit race for the Brain Tumor Center.

It was a Saturday, only five days beyond her last radiation treatment and nine past the most recent chemo dose. She was supposed to be exhausted. Yet there she was, now and then breaking into short jogs as we wove through a crowd that took over a parking lot just below the Duke Medical School, most of them in running shoes, some in small groups of matching T-shirts. Bright orange ones here, light blue there. "Casey's Warriors" on one set. "We Love You, Robert" on another, beneath a silkscreened picture of a smiling man astride a motorcycle. Folding tables under canopies of red and blue and green contained pamphlets and still more T-shirts, while at the back of the lot, an inflatable bouncy castle began to take form. Parents were helping children pin on race numbers, while a few more-serious-looking competitors ran warm-up intervals on the road just beyond. The joyous energy of a normal race was present amid a palpable weight.

A white banner crossed the road, emblazoned with "Angels Among Us" in blue scripted letters. A tall and angular man wearing one of the race T-shirts and holding a megaphone stood on a temporary platform to which one end of the banner was tied. Fifteen years beyond his own glioblastoma diagnosis, he was a living emblem of hope. He exhorted the crowd to carry the spirits of lost loved ones with them, to live each day with energy and grace. And to never, ever give up. Then he began the start countdown. There were tears and shouts alike as the throng's prior random movement coalesced into forward progress.

The crowd was sprinkled with people who came to be with Diana on this day. Once again, friends from nearly every corner of the country, most of them a planned surprise. The night before, all of us had crammed into the living room of our house as they talked of running the race with her. Of strength in numbers out on the course.

We can run as one big group—whatever pace feels okay to you.

To that, she just smiled. I knew why and was proven right when she disappeared ahead of me and most everyone else as the race began. She'd done the same thing years ago in a Thanksgiving fun run near her childhood home and in the lone marathon we ran together. Each time flashing me a smile before saying, *No friends on race day.*

Cancer was not taking that from her.

When I saw her again, she was in a group of those friends, her face glowing and her smile as big as ever. One friend broke off to intercept me as I approached, still breathing heavily from my own run.

Un-fucking-believable. She flew. I couldn't catch her at the end.

I stood and watched her from a slight distance. She hugged one member of the group, then broke free and threw her head back in an eruption of laughter. Her face remained flushed, looking just as it had when we walked down the final hill of our first runs in Costa Rica more than a decade before. She had stopped back then in childlike wonder to watch a bright green tree frog with brilliant red eyes cling to the branch of an understory palm, while droplets of a soft rain fell from the brim of her hat. A little earlier on that same run, we rounded a corner to find a capuchin monkey uncharacteristically on the ground and blocking our path. The monkey raced up the nearest tree and out onto a branch that overhung the trail before shaking it and screeching in anger. She had laughed in delight.

After another such run on a return trip, we paddled kayaks from the river mouth behind the hotel out into the waves just beyond. We rode those waves for nearly an hour, often falling from the bulky, hard-to-steer sea kayaks into the foaming water below before clambering back on to do it all over again. At one point we floated beside each other, waiting for another set, and she pointed to a long dark object that was also on the ocean's surface, just past the end of where the waves broke.

Why does that log seem to be moving into the river mouth current?

I looked and took a moment before answering.

Ah. Yeah. It's not a log. It's a crocodile.

She stared at me for several seconds with a look I'd already come to know well. When she spoke, the words were slow and measured.

You realize it has been there the whole time, right?

Um. Yeah. He's out here a lot. Never bothers us.

At that she simply shook her head and then caught the next wave, surfing it right toward the croc before paddling back up the river. I followed suit and caught her as we both reached the dock, where she hopped out of her boat and left it floating slowly back down the river before turning to me with a purposeful grin.

I think you can put my kayak away. You know, in case there's a crocodile in the boathouse.

Now I pushed through the assembled throng of friends, embracing her in my own hug. I asked her how she felt. Instead of answering, she said, *I think I did pretty well! Can we stick around for the awards ceremony?*

Soon she posed for a picture with an incredulous Henry, on the heels of being announced as the winner of her division. She tore off her running hat with a flourish when she crossed the stage to accept her award, turning the scarred side of her head toward the crowd for all to see. They cheered. Henry stood with his arm around her shoulder, cast a look in her direction, and shook his head. For once he didn't have a stream of words. Only, *How did you manage to do that?*

I don't know! It was hard and the whole thing is kind of a blur. I was so dizzy that I nearly ran into a tree.

She walked off to rejoin some of our friends while I turned to Henry.

Is this normal? Do all of your patients go win races and basically ignore advice to take it easy?

He laughed and shook his head.

No. But you'd be amazed what this can bring out of some people.

What do you think defines the people who do best with it?

I don't know. Some are just those I'm-tougher-than-you types, but in truth, they tend to have a combat mindset that I'm not sure is so helpful in the end. She seems different. Like she's somehow at ease.

You know she basically hasn't stopped doing science and is still planning to go to Antarctica.

Henry laughed again and then scratched his mottled beard.

Good. Maybe that's why she's doing so well.

Only hours later, I sat by an airplane window while Diana was on the aisle. Neva happily colored between us with her feet propped up on a polar-bear-adorned suitcase. We were bound for New Orleans, having gone straight from the race to the airport, and as we began our descent, I could see the broad expanse of the city itself and the serpentine southern river border that lay just beyond. It was a largely foreign land for me, a home for her. We'd come for Jazz Fest, to see her favorite group—the Rebirth Brass Band—take the stage and remind us in a sweltering heat that no matter what comes to pass, we still have choices that belong only to us. "Do Whatcha Wanna." Soon we emerged into the heavy air, collected our rental car, and sped along I-10 beside Metairie's stained canals and haphazard construction, to arrive at a hotel in the heart of the French Quarter.

The choice surprised me because she loved the forgotten edges of New Orleans. On our one prior trip, we stayed in a shotgun apartment tucked into a far corner of the Marigny, so that we could be closer to the Ninth Ward, its hallways adorned with grayscale acrylics of nude men and memorabilia of the Krewe of Petronius. We shucked and ate a mountain of crawdads in a ramshackle joint south of Saint Claude. When we went to the Quarter, it was almost an afterthought, as though a box needed to be checked. So, I asked.

Why did you want to stay here?

I don't know. I just...

She paused and looked at me, and I saw a tightness come into her face. A hint of pleading in her eyes. I wondered if she had to be in the city but in her current state couldn't face the additional pain of what had happened to the places she loved most. So I dropped it and redirected.

It'll be fun though—let's go show Neva the circus.

Dada, we're going to the circus?

I laughed.

No, honey, it's just an expression. This town is kinda crazy.

The next morning brought an almost eerie quiet after the prior night's heat and noise. We angled toward the river and found Jackson Square bathed in a diaphanous mist. We passed through the square as vendors began to set up tables for the day and crossed Decatur. Diana wanted beignets from Café Du Monde.

Again, I wondered why she steered us to a place on every tourist's list. But I didn't ask. The beignets were good and the part where she and Neva started smearing powdered sugar over each other's faces, even better. We sat there for more than an hour, and by the time we walked out, the Quarter was waking up. Neva successfully lobbied for a stuffed bear and a pink handheld fan, each of them emblazoned with the name of the city, then broke the fan twenty minutes later. We went back to replace it. This was not a vacation of restraint.

When it was time to go to Jazz Fest, we wound out of the Quarter to the northwest, along the edge of City Park, and drove a few laps north of the Fair Grounds in hopes of finding a parking place that wasn't too far off. Success. As we walked toward the venue, Neva swung between us, kicking her feet high each time until all our hands became bathed in sweat. Jesus God the heat here, and it was still spring. I said as much, and her pointed retort was quick but delivered with her shit-eating grin.

Waaaaah!

That smile always went straight through me, but now when it shone beneath her polka-dotted bandanna, it hit like shrapnel, piercing my heart and gut alike. No matter how hard I tried, I could never step far from the specter of what lay ahead in our lives. There were times when I envied how she seemed to be able to set it all aside so easily. Her mental toughness that was, I now believe, at least partly a direct product of her relentless curiosity.

There are good reasons to believe that this curiosity was her secret power. In Norman Doidge's book *The Brain that Changes Itself*, he describes the concept of neuroplasticity in this context—evidence that for those who are constantly seeking new information, the brain builds upon itself, laying in an ever-greater capacity to explore new ideas and the world. Doidge also talks about how dopamine-based "rewards" come not just from getting an answer but often even more strongly from the anticipation that it might arrive. In other words, the habit of curiosity trains the brain to respond positively to simply seeking answers... without necessarily having to know them. And brain scan data suggest that those who do this regularly are less reactive to stress. This was Diana.

Neuroanatomist Jill Bolte Taylor is a living demonstration of neuroplasticity. As I listened to her discuss her remarkable story on a podcast recently, I thought of Diana. Taylor has devoted her life to studying the brain, but at thirty-seven, she suffered a cerebral hemorrhage that essentially shut the left hemisphere of hers down. She described it as having reduced her to "an infant in a woman's body," and as chronicled in her book *My Stroke of Insight*, it took about eight years for her to return to complete function. But return she did. Her brain adapted, healed, and rebuilt itself. What stood out to me, though, was this. When asked if she was afraid as she watched her cognitive abilities decline in the hours following the hemorrhage, she replied no, and that instead, as a scientist, she was "one hundred percent fascinated." She goes on to state how she's certain her scientific curiosity was essential to her recovery. In her words, I saw my wife.

We funneled into a queue that pushed us through a double set of turnstiles and then along a wider dusty track, fresh faces milling in, sunburned and bleary-eyed ones headed out. The music grew louder as we walked past the line of buildings and out into the main crowds. Rebirth wouldn't begin for another half hour, so we wove between bodies and blankets and folding chairs to reach the food stalls. I lifted Neva to my shoulders after she was nearly trampled by a pack of beer-spilling frat boys. Her eyes big, cheeks flushed. Nothing a heaping basket of fries couldn't fix.

Then it was to the field in front of the stage. We found a spot beside a couple in their sixties, maybe even older, both of them exuding a casual elegance against the general drunkenness of the crowd. Each had large freckles beneath gray hair, his tightly wound and only slightly receding, hers in bulky dreads and loosely bound between her shoulder blades. They sat beside each other, legs extended and his hand resting lightly upon hers. As we approached, they smiled broadly, and the man patted the ground beside him in welcome before turning to look at his wife with unabashed adoration. The moment relit my pain.

But then Rebirth began and the couple swayed and Neva twirled beneath Diana's hand and the pain within me dissolved away. When the song we traveled to hear burst forth, we all danced and danced and danced as though not another soul was on the surrounding fields, flowing through each other like intersecting threads of a river eddy that you know must eventually release the water it contains and yet feels as though it will never end.

CHAPTER FOURTEEN

The heavy air of a Piedmont summer began to tease its imminent arrival. A shirt back moist after yet another walk to the cancer center. A new curl to her hair where it once had emerged from the bandanna's border. A suddenly laden sky opening in the early evening, sheets of rain sculpting mini-watersheds through the mulch before us as we sat quietly on the covered porch. The bandanna was now gone, replaced by a jauntily butch haircut she wore with pride. The scar remained obvious and yet no longer stark. Her eyes were bright and her jawline relaxed as she watched the rain pulse onto Markham Ave.

Rain brought with it memories of the Costa Rican forests. Of how the parched first months of the calendar year send the animals into hiding and turn parts of the forest bare. Of how exposed soil in the sun-bleached pasturelands will crack while skeletal cows seek relief beneath whatever trees are left behind. And of how the forest bursts with new leaves and bright flowers after the spring rains come. At this time of year, the eponymous *mayo* trees would be in bloom, creating a sense that each distant ridge of green had been splattered with yellow paint.

Over the years these forests had doled out their share of surprises. One of those came when trying to prove something I was already convinced was true: that they depended on the nearby ocean for critical parts of their nutrition. That trip started poorly, with me running over my graduate student Erica's foot. A combination of soft clay and a hard boot rendered the incident no more than a close call, but a pattern began when my other student, Carl, whose thesis was wrapped around this particular campaign, decided he didn't give a damn about the nearly crushed foot. He was overly stressed about getting the samples he needed; she was there to help but increasingly unmotivated to do so given his behavior. They simmered and sniped. And I barely noticed, because Diana rounded out our foursome and we were still in the honeymoon stage of our relationship.

I shirked my mentoring duties, even fueling the conflict by repeatedly pairing them for one field task while Diana and I would do another. The days passed and their war escalated while the two of us floated through days and nights around the Osa, taking soil and leaves from a pocket of resplendent forest, moving on to another the next day. In one remote forest near the eastern shore of the Golfo Dulce, we sat against an inky scarp in a narrow gap between the rock and a waterfall, ducking our heads in and out of the cascade. To our right, a pair of black and neon green poison dart frogs slowly climbed the wall in staccato hops. High above, a canopy of multihued green completed the borders of a baptismal world in which the water carried rivulets of red earth off our sodden clothing, while fusing us together and to the rocky remnants of a seventy-million-year-old sea.

We left the waterfall reluctantly, grabbing our packs and shovels and beginning the climb to a nearby ridge that was our destination for the day. The slope was steep enough that you could easily reach out and touch the ground before you, and we often needed to grab the bases of sapling trees as handholds. At one pull and step, a mottled brown form slipped quickly to my right, repeating black triangles and a diamond head visible before it disappeared beneath a fallen tree. I jumped back.

Terciopelo. Let's go this way.

Diana looked at me quizzically but said nothing as we continued to climb. By the top, the sun was breaking through both the morning clouds and the thin strip of forest canopy still above us on the narrow ridge. She dropped her shovel and pack and looked at me.

What did you say back there?

Terciopelo. Fer-de-lance.

A pause.

Next time you use a fucking Spanish name I don't know for a snake that might kill me, I'm going to hit you in the head with this shovel.

She was smiling. Sort of. I held up my hands in surrender. Then we each took up shovels and began to dig. Perhaps thirty minutes later we had a hole more than three feet down into the red clay, with most of one wall of the soil pit scraped smooth and flat so that it was ready to take the samples we needed. She lay prone with her head out of sight while finishing the wall polish near the bottom, legs apart and the toes of her rubber boots dug in to brace herself from falling completely into the hole. When I leaned over to assess her progress, she flung a trowel of soil into my face.

Watch out. Terciopelo.

A cackle from below.

We finished the pit sampling and then moved on to another round of leaf hunting along the ridge top. Diana turned out to have a talent for shooting them down, so she'd fire away with the shotgun while I ran around the forest trying to catch the leaves as they fell. At one point I smacked into one of the thorn-covered palm trees, leaving my right arm looking like a dog after a porcupine encounter. I dug out a Leatherman tool and then sat on my pack while Diana pulled the thorns free, giving me gentle rounds of shit all the while.

We were digging the pits and collecting the leaves to find out how these forests got their calcium. Just like us, trees need a laundry list of different elements to run their lives, with a few in fairly high concentrations. They get their carbon from the air via photosynthesis and, ultimately, nitrogen—an essential component of proteins—comes from the air too. But just about everything else they need, including that calcium, comes from rocks. As those rocks slowly weather away under the influence of heat and rain and the intrusion of plant roots, they give up the bounty they contain. And in most places around the world, there's enough remnant rock around to keep everybody happy.

Not so in many tropical forests. Many of them perch upon some of the oldest soils on Earth, where the rocks are long gone, so they need other sources of nutrition. As I tried to lay out for Manuel all those years ago in Brazil, forests in the Amazon depend on dust from the Sahara. Forests on the island of Kauai depend on rainwater that comes from the nearby sea for their calcium and magnesium and on dust from the distant Gobi Desert for their phosphorus. One more way the world is astonishingly connected across vast amounts of time and space. And here on the Osa, I was sure the answer would be: just like Kauai. The soils were old, the rainfall was high, and the ocean was next door. There were no rocks to be found anywhere in those soils. Sometimes you do research just to reinforce what you are certain you know.

And sometimes you are delightfully wrong. We learned that the Osa had some of the highest uplift rates in the world, with the ridge where Diana threw soil in my face and all those around it moving skyward about half an inch each year. That doesn't sound like much, but as it turned out, it was enough to keep injecting new rock from below that provided nearly all the nutrition the forests needed.

Results that went against all expectations and evidence were some of my favorite moments in science. Now I wanted to believe that my wife could become one of those surprises, that somehow her tumors would not follow standard expectations. And though daunting, the odds were *not* zero. About 5 percent of glioblastoma patients make it past five years. Diana was giving me, and others, plenty of reasons to hope that she would be part of that select group.

Brain tumors are a particularly cruel form of cancer because they can rob a person of their humanity. In Neva's case, minor tumor growth could have taken away her sight before then taking away her ability to know when she was hungry or thirsty. There are children with craniopharyngiomas who live a heartbreaking existence in which they cannot see, while they relentlessly eat themselves into a cascade of new health challenges, all before their lives are cut short. And for Diana, a glioblastoma could take a wrecking ball to any part of her brain and thus any part of who she was. The horror of these brain tumors was not just the loss of life but the loss of living.

But Diana refused to stop living. In ways that were sometimes flat-out mind-boggling. As she did so, day after day, I began to see how her choices flowed well beyond her own body. And I realized that while I was obsessively searching and hoping for her miracle cure, for the most part, she was not. She was just moving forward, as fully as she ever did.

Christians often find comfort in the promise of what comes after our mortal lives, feeling their allegiance to God will carry them through. *Yea, though I walk through the valley of the shadow of death, I will fear no evil.* For many, there is enduring power in these beliefs, and I had moments during Diana's illness when I wished they were oncs I shared. But Diana's ease seemed to flow from her relationship with science. It let her live in the present, buoyed by a fascination with the unknowns of the natural world, rather than dragged down by the unknowns of her future. Sure, she wanted a cure. But she wasn't consumed by it. And in that she was showing me how the practice of science can soothe the soul.

She seemed to know that if we pin all of our hopes on the discovery of those cures, we're once again getting trapped in a game of chance we will likely lose. Science teaches us that yes, miracles are possible, but also that there are limits to our control, whether that be curing cancer or shaping behavior or even simply building a bridge. If you approach science with the mindset that success is only defined by that miraculous, breakthrough answer, you're missing a critical part of its power. If we let it, science can also teach us about limits and acceptance, without which we can never find peace.

If Diana *was* to find a miracle cure, it was most likely to arise from the experimental vaccine trial. So on a humid June morning, we drove the familiar route to the cancer center, where she once again donned a thin hospital gown. I'd come to hate these gowns. They seem designed to strip away one's humanity right when you need it most. This one was a faded blue and white, and the right sleeve was pushed up a bit to allow for a pair of large needles in her arm that were connected to clear tubes, with flowing blood visible in each.

The blood exiting her body disappeared into a boxy machine at her bedside and then pooled, darker, in a collecting vessel below. The second tube returned most—but not all—of the blood to her body. She lay there for almost two hours, spending most of it quiet, eyes closed, her breaths long and even. Meanwhile, I stared at the container of blood because it held my outsized hopes of beating glioblastoma's awful odds.

The bedside machine was filtering her blood to concentrate the cells within. Soon the blood would undergo a series of further separation treatments to pull out a vial of immune cells. When viewed under high magnification, these cells have astonishing beauty. They look nothing like the classic cell diagram you saw in your high school biology text but are instead constructed of intricate folds and often radiating gossamer branches. At times the center will appear as a translucent blue of infinite depth. They are more nebulae than cells. Which felt right.

Known as dendritic cells, they are scouts of the immune system, largely present in the parts of our body that feel the assaults of the outside world. Skin, nose, stomach, lungs. They are on the lookout for threats that can make us sick, known in immunology terms as antigens. When they find one, they take it to a lymph node and hand it off to T and B cells, which then unleash the remarkable immune responses our bodies can mount.

Despite the fact that in blood dendritic cells are not yet fully formed, it was the only way to collect a group of them.

Post-harvest, they would be coaxed to mature further in specialized lab chambers and then hopefully—oh God, hopefully—tricked to kick off an immune response against Diana's tumor. They would be exposed to antigens formed from glioblastoma cells and then turned into a personalized vaccine just for her. If it all worked, the vaccine would trigger a targeted attack within her brain.

It was but one of dozens of strategies in a frontier of new hope for cancer treatment known as immunotherapy. Boiled down, all of them were branches of a common idea tree: make our immune systems recognize cancer cells for the threats they are, then let the body's innate defense systems take them out. It is an idea of exceptional promise, for it harnesses the unmatchable adaptive power of our own bodies, creating the potential of finely tuned responses to each person's tumor. For the most part, all other cancer treatments are cruder instruments, and therein lies their limitation.

The vaccine plan was seductive, an anchor for months now. Early trials looked good: in one, half of the patients were still alive beyond three years, compared to an average survival time on a standard treatment of about a year. As such it was not a magic pill, but perhaps it was a step to a subsequent breakthrough, one that embodied our hopes. She was young, strong, and had joked for years about her "badass immune system," which seemed to keep her from ever getting sick.

Yet we knew the odds remained long because glioblastoma had its own daunting arsenal. It began with a defense of the brain itself, known as the blood-brain barrier. This firewall of dense cells is there as a blockade against infection and other assaults, but that can bar cancer treatments from entering where they must. And for a specialized vaccine, we also knew the brain's immune system was unique and might not respond. Then there was the tumor itself, notorious for its ability to ward off immune responses or any other treatment. Glioblastoma is an empusa, changing form again and again as it consumes and shoves aside the very essence of what makes us human. Not that many other cancers don't do extraordinary damage in their own right, but we had learned from Henry that glioblastomas were particularly good at evolving and adapting quickly in the face of treatment. You might slow them down for a little while, but they'd still be there, gearing up for their next chance to take off.

Still, maybe this time the tumors would fail.

As I watched her breathe steadily through the cell-extraction procedure, the fusion of science and love was lain bare before me: the products of uncountable hours of hopes and dreams and desires of those seeking cures for life-shattering diseases were literally flowing into my wife's veins. It was a reminder of how there is enormous power in bringing emotion and humanity to science.

The best scientists I know are deeply invested in their work for reasons that are more fundamental than career ambition. They love it, and in that very attachment, their emotions are woven throughout their practice of science. The tropes of the automaton scientist or the need to remove emotion to ensure rigor are frankly ridiculous. My first exposure to research was as an undergraduate, and my adviser—the evolutionary biologist Paul Ewald—would visibly light up and almost seem to burst with happiness when he talked about what evolutionary theory could tell us about our daily lives. A couple of years later, I was on the side of a Hawaiian volcano with my PhD adviser, the ecologist Peter Vitousek. He smiled broadly as he pointed at a bright red bird with a long and curved orange beak that hopped among the branches of the ohia tree in front of us.

'I'iwi, Peter said. Pronouncing it ee-ee-vee. *We're lucky to see it.*

He kept smiling as he watched it fly off. Then he talked about the history of the spot on which we stood and how that helped explain many of the things we could see. Finally, I remember him saying, *I can't believe I get paid to do this.*

Like me, Peter had grown up in Hawaii, and that day began to demonstrate how unraveling the secrets of its natural landscapes was almost basal in him. His need to know more about the islands arose directly from his deep love for them.

Pick just about any famous scientific discovery you can think of, and behind it will almost certainly be someone who was consumed by passion for their subject or for a solution they desperately wanted to find. Charles Darwin was no fan of his early formal schooling, but he couldn't get enough of learning about the natural world. When things went well, he was reportedly full of joy. And when they didn't, he could slip into a funk. One day in the midst of his famous voyage, he wrote the following to Charles Lyell:

> But I am very poorly today and very stupid and hate everybody and everything. One lives only to make blunders. I am going to write a little book for Murray on orchids but today I hate them worse than everything.

But soon enough, his letters regain a sense of optimism, arising directly from his observations of the wonders around him.

Nobel Prize winner Barbara McClintock is famous for saying: *I know my corn plants intimately, and I find it a great pleasure to know them.* Long before my family's life was turned upside down and my own lens on science would change, Carl Sagan's boundless passion for the intersection of science and society had already shown me the way. In his book *The Demon-Haunted World: Science as a Candle in the Dark*, he wrote:

> Science is not only compatible with spirituality; it is a profound source of spirituality.

I wasn't all the way there yet. But I was regaining my faith in science. It had been tested on the heels of Diana's diagnosis, but the vaccine Diana would get was both a path to a miracle and only possible in the first place because of decades of painstaking, passion-filled efforts in which scientists I would never know just kept going in the face of innumerable setbacks. When we got home that night, I felt a fresh round of confidence, and the *fuck you, tumors* I whispered dozens of times each week had a newfound arrogance. This time Diana heard me say it.

What's that?

I said fuck you, tumors.

She looked at me for a moment.

Why?

Because I don't think they deserve a more polite goodbye.

CHAPTER FIFTEEN

Five years after Neva was diagnosed, I participated in a documentary film series called *Let Science Speak*. The filmmakers pitched it as an attempt to "show scientists as people, to show their real stories, to focus on their hearts as well as their minds." They wanted me to tell Diana's and Neva's stories and to explore what it was like for a pair of scientists to confront the cancers in our family. The brainchild behind the project was Christine Arena, a former PR firm executive who quit the business with a vow to use the tools of her trade for societal good. She was tired of how frequently she saw science and scientists being both ignored and unrealistically portrayed. And she's right: the scientists we see on screen rarely mirror the reality of the field I know.

At one point a few years back, scientists on Twitter had a lot of fun mocking Hollywood portrayals of the discipline. The string of posts ran the gamut from the impossibly smart scientist who can solve anything, to the nerd who can't relate to a real person, to the inevitable framing of nearly every woman scientist as a sex symbol who cannot, in the end, save the day without the help of some guy. And perhaps the most common theme was: the scientist who nobody believes but who will of course find the miracle solution to the problem at hand. All of this reinforces a public relationship with science that frequently views it as only a source of answers and solutions and perceives it as a failure when those solutions don't arrive.

But here's what wasn't in that Twitter string or anywhere on-screen that I've seen: acceptance. Not in a Hollywood-esque screaming-into-the-void fashion, where the scientist is desperately trying to get society to see the imminent danger, but instead showing science as a path to finding ease and even joy with the world as it is. Perhaps Jeff Goldblum in the original *Jurassic Park* veers closest, when he simply says: *Life finds a way.*

Acceptance in a scientific sense doesn't mean giving up if life is throwing difficulties at you. It just means that science has a way, if we let it, of training our minds to see nuance and unpredictability and a range of future possibilities at nearly every juncture, some good, some maybe not. When it comes to something like cancer, too often it becomes framed as combat, which in its very definition means creating more stress, in this case at a time when that's absolutely not what is needed. Because that stress makes us less effective at both healing and at solving the problems that might be in front of us. Diana chose a different path—one where someone can still fight like hell for the outcome they want but recognizes that remaining at ease is part of how you tip the odds in your favor. As I learned over the final weeks of summer, part of this approach to life, including its inevitable end, means accepting that those odds still might not work out.

July brought another good MRI that showed no progression of the tumors in Diana's brain, so we decided to take a trip back to Boulder. There she felt healthy and strong, so much so that she insisted on renting mountain bikes and going for ride after ride in old haunts. At one point she lost control of the bike and fell but waved aside her bloody leg despite her medications creating the risk of an inability to clot. She simply tore up the shirt she was wearing, wrapped the leg, and carried on. When I started to protest, she said, *What? I have a sports bra on. And there are worse ways to go than bleeding out right here.*

That night we sat with old friends and ate and drank and laughed and even cried a little. As we left the dinner with Neva in my arms, a classic Boulder sunset turned wisps of cirrus clouds aflame above the angular shadows of the Flatirons. As if by instinct, I turned away from the route back to our hotel and drove through the South Boulder neighborhoods up toward the parking lot of the National Center for Atmospheric Research. We sat in the car as Neva drifted off and the darkness emerged, the brightest of stars increasingly visible above the city lights below. Diana steered our conversation to the future.

I think we need to come back to Colorado. Or if I don't make it, you and Neva need to. I think we should try to get a place here if we can.

I sat in silence for a time, not wanting to absorb the latter as a possibility. But knowing deep down that she was right. Eventually I managed a quick, *Okay.*

She pressed on.

If I don't make it before we return, I want you to do a Call and Gus.

A what?

Call and Gus. Lonesome Dove.

She pointed out the car window in the general direction of our old home.

I want you to bury me in that cemetery right over there. The one behind NIST at the bottom of the hill.

In the infamous final chapters of McMurtry's novel, a dying Gus McCrae tells his lifelong friend that he wants to be laid to rest two thousand miles away from the eastern Montana town in which he passes. Call is not happy with the conversation or the details of Gus's postmortem plans, and I felt as resistant and hidebound as his character, but I asked:

I don't think we need to talk about this because you're going to make it. But... why that cemetery?

Because the upper corner of it is next to my test run.

Your what?

The trail I always ran to see if I was really in shape or not. Goes straight up to the Mesa Trail. Feels great when I'm fit. Hurts like hell when I'm not.

More silence.

Promise me you'll do it. It's like Gus said—this is for you too. And you don't have to drag me behind a horse, so don't bitch about it.

I couldn't help a short laugh.

Okay, okay! Not gonna happen... but I'll do it.

Later, as we lay beside each other, she said she had no intention of being in that cemetery for many years to come. That she was just making contingency plans. But too soon those plans came roaring back.

In August, it was time once more to give Neva a final kiss as she disappeared into the MRI room, her stuffed giraffe clutched tightly, her body covered only by the little gown. As before, we stood beside her soon after she woke, helped her take her first sips of water and eat a package of Goldfish crackers, and urged her to be brave and relax as the IV was removed. Then we walked her slowly out of the ward and headed for the ice cream stop that had become tradition. So too had been the good news that followed when her scans were read.

Not this time. When the oncologist suggested his nurse take Neva down the hall to play, we knew what was coming. The moment she was gone, I asked, *How bad is it?*

He pulled up the black-and-white images we knew too well before saying it was nothing to panic about. He told us the new growth was small and that we should simply increase the frequency of her MRIs and watch it for now. That there was no reason to move to any treatment plan yet because the growth posed no imminent threats. But we all knew what this meant. When a craniopharyngioma begins to grow, it doesn't stop. Sooner or later, something would have to be done. And so we found ourselves once again having to shift under the weight of another blow while trying to assure Neva that everything was okay.

That became monumentally harder only two days later, when we sat with Henry and listened to him tell us that both of Diana's tumors were now growing. He was especially concerned about the one in the brain stem they had been unable to surgically remove. That one had grown significantly in only a few weeks. He pulled no punches, but he tried to balance it with a reminder that she had just received the first round of the vaccine.

Thank God we got that rolling. It has something to work on now.

We scarcely heard him as we stayed rooted to our chairs, silently locked on to each other's eyes.

That night we simply sat together as a family, all of us tight to each other on the couch. We read to Neva and held her as she cried, her body sensing the change. When we cried a little too, we didn't try to tell her everything was fine. We just said life was hard sometimes, and our love would get us through. It felt like we were being honest with her for the first time.

The following day was the first in months that Diana didn't spend a moment on science. She took our daughter to the park, played nearly an hour on the swings, came home, colored, read, and then let herself slip into something she never did: a long nap. Neva slept with her, curled into her mother's stomach and arms. I spent much of their nap sitting across the room in an armchair just watching them sleep.

Yet by the next day Diana was back to her old self. She announced a need to go to the office that morning and a desire to still attend a good friend's wedding in South Carolina the following week. We left Neva with the Chamberses and did just that, spending the weekend surrounded by Diana's childhood friends, swimming in warm and gentle waves, dancing at a reception where she was constantly taken from me to fall into the arms of another blast from the past. It was all beautifully distracting, broken only slightly by her admission that she felt like she was losing some control of her right arm. And that she seemed to be having a hard time finding some of the words she wanted to say.

This isn't unusual, Minh said when Diana passed on these new developments. *But it may get worse, so it's a good idea to start some countertherapy right away. For both speech and motor control. I can set it up.*

That meant an iPad loaded with apps for communication and motor exercises and a series of appointments at yet another branch of the Duke medical complex, where she endured a battery of tests. She got fed up and asked to go home.

I don't want to do this right now. It's not that bad—hell, I'm still running and typing fine on my computer.

I looked at her with concern.

Are you sure? Minh said it was good to get out in front of it.

Yeah, I know. But it's only dragging me down. I want to do stuff that makes me feel good.

She cemented those words by running the final leg of an annual triathlon that was held at Duke's Marine Lab. We formed a last-minute team with a swimmer, while I rode the bike leg. I touched her hand at the line that was both start and finish, along a section of Pivers Island Road that separates the brown sidings of the Marine Lab buildings from the bay. As she took off along the path I'd just completed, she had one runner to catch. She did so in the final half mile, coming across the finish line alone to give us the team win and running hand in hand with Neva for the final few yards. At the banquet that followed, she confessed that she could barely see for portions of the final mile. But the satisfaction on her face was clear.

Back home we tried to reclaim old routines. She insisted on spending her days in the office, still postponing the speech and physical therapy sessions. After her triathlon performance, it was hard to argue the point, even as I noticed her shake that right hand repeatedly. As she'd done the year before, she dug into options for Neva's treatment to come, contacted doctors she'd met in Germany, and called Todd. Who gave us a ray of unexpected hope.

I agree that for now it's best to just wait. But if and when the time comes, I don't think you're stuck with radiation as the primary option. Usually that would be the recommendation, because second surgeries for cranios tend to be a much bigger deal. And I'm not saying surgery will definitely be the right choice. But in Neva's case, her healing from that first one was so clean that I think I could go in through the nose again without much greater risk.

It's odd to have *brain surgery might be an option* feel like great news, but it meant she had another chance to get the entire tumor out and to do so without the collateral risks of radiation. It also meant that when the time came and if surgery was the call, we had to find a way for Todd to do it. Our conversation about returning to Boulder was reignited.

The next morning I walked Diana to her office from the parking lot that separated our buildings, then headed to mine. Within an hour, I called to ask if she wanted some coffee. It was a thinly veiled pretense to check in.

Sure. Can you bring me one?

I'll be right over. You feeling okay?

Yeah, though super tired. Not sure why. Coffee will be good.

By the time I reached her office, she was lying on her faux-leather chaise with her eyes closed. At first I thought she'd decided to take a nap, but her face was screwed up tight and she was clutching her right forearm with her left hand. I set the coffees down and rushed to her side.

Honey, what's wrong?

I don't know. My arm hurts more and also feels really numb. So does my leg.

Okay. Shit. Do you want me to call Henry or Minh?

I don't know. Maybe.

The first waves began as I dug for the phone, not as violent as before but unrelenting, her arm and leg alike shaking uncontrollably as her eyes widened in fear and her left hand gripped the side of the chaise. I dialed 911 instead, the call routing me to an internal Duke emergency switchboard.

My wife has a brain tumor and is having a seizure right now. In her office in the biology building.

What is her office number? I'll get a team dispatched right away. Also, who is her doctor here?

I gave the information and hung up, called Henry, called Minh, left messages for both. Meanwhile, Diana continued to shake, and I tried to keep her calm.

Remember last time, honey. It will be fine. These happen. Just try to relax, and help will be here soon.

It wasn't. Twenty minutes on and still no team had arrived, as both of us became increasingly anxious while Diana was still in seizure. Pulse after pulse, milder at times and borderline violent at others, never ceasing. I called 911 back two more times.

Where the hell are they?

Finally, a pair of young men entered the office and then announced they were not paramedics but part of Duke security and were there to help as they could until the EMTs arrived. I couldn't help yelling out, *Are you fucking kidding me?*

But the two were kind and soon joined by the paramedics themselves, both of whom radiated competence. One began prepping a pair of syringes and vials while the other sat beside Diana and asked her questions in a calm voice.

Diana, can you hear me?

A nod, then a weak *yes.*

I'm sure you know this, but you're having a seizure. The only thing unusual about it is how long it's lasting.

He turned to me.

You said half an hour now?

Yes.

He turned back.

We spoke with Dr. Friedman on our way in here, and he wants us to have you taken into the hospital. Okay?

She nodded.

First, we are going to give you some meds that should help stop this. We'll give it a few minutes to work and then take you in.

Another nod.

The seizure finally stopped, and the paramedics wheeled her down the hallways of the biology building toward a waiting ambulance. She was secured to the gurney by two blazing yellow straps. Exhausted, eyes closed. At one point we passed a group of graduate students we both knew, all of them pressing their bodies against the wall so she could pass. They looked at us with a mixture of shock and sadness and uncertainty about what to say. I managed a halfhearted, *It's okay.*

But it wasn't. As I sat beside her in the ambulance for the short ride to the Duke ER, I had the awful feeling this marked a point of no return. Only two weeks later, her right arm was almost useless, her right leg weakening, her speech noticeably more labored. I clung to a hope that it was all a product of immune-driven swelling in the brain that could arise from her vaccine. A known and temporary side effect. She seemed to know otherwise and began to force conversations about what would come next with newfound urgency.

Honey, I don't think I'm going to get better.

Please don't say that. Don't give up.

Her eyes flashed before she said, *When have you ever known me to give up?*

I'm sorry. You're right. But this could just be your vaccine response.

I know. I know. But we have to accept. The possibility. It's not.

Her words came slowly, with pauses between them, as though she had to forcibly extract each one from within and push it out through a viscous path.

I need. You. To promise me. That you and Neva. Will move on.

I fought back tears, and though I knew the answer already, I couldn't help but ask, *What do you mean?*

I mean. You can't let. Me. Me.

Long pause.

Dying. Keep you from. Living. Finding someone new. To love.

I don't want to talk about that.

I know. But. You will... need... it. And so. Will Neva. Promise. Me.

I nodded slightly as I looked at her through wet eyes. Then, as ever, she lifted us both. She pointed with her good hand to a small white box on the mantel, wrapped with a gray ribbon that had a repeating series of paw prints. The box contained the ashes of a favorite dog that had passed away when Neva was an infant. She looked at me with a new light in her eyes.

If you don't. Do this. I'll come back. And. Sit in a box... beside... my... dead pit bull. And haunt you forever.

I cried and laughed and hugged her and told her she was a giant pain in the ass.

CHAPTER SIXTEEN

We hear it all the time: correlation is not causation. It's one of those scientific principles that has spilled over from the field itself into our general lexicon. And yet we ignore it all the time. Humans are hardwired to infer cause from pattern, and even scientists are not immune to the traps it can spring. Thinking like a scientist means you are supposed to observe and seek out patterns and then try to figure out what caused them. But in that second part, you have to be very careful.

Why? Because we're all human, and our evolutionary roots set us up not only to see those patterns but to preferentially gather information that supports our preconceived notions, often unconsciously. That so-called confirmation bias is woven through us all, which is why some of the foundations of science went through backflips to guard against erroneous attributions of cause. The philosopher David Hume famously proclaimed that you can't prove causation. Ever. Karl Popper agreed that we can't prove anything, only disprove it, and as a result the statistical standards in science are set up to do a very weird thing: falsify (to a certain level of odds) that data sets are *not* correlated. Huh? Yeah. Sometimes thinking like a scientist is super confusing.

Leave it to that highly respected institution, the Church of the Flying Spaghetti Monster, to demonstrate the trap effectively. Want to understand why the climate is warming? The Pastafarians say look no further than the decline of piracy. As the number of pirates in the world steadily dropped, the world's average temperature inexorably rose. Coincidence? THEY THINK NOT.

Okay, all very funny. Except when it's not. Falling into cause-and-effect traps can shape entire societies, often for the worse. COVID's recent examples surrounding vaccine skepticism and conspiracy theories about viral origins are only the latest in a long line of deadly misdirections, a grim and sometimes darkly hilarious list that even includes people once believing lice were beneficial to our health—simply because sick people with fevers didn't have them. Turns out that's because lice don't like an overheated body any more than we do.

Debates over correlation and causation have become rampant in climate science as well. One of the most famous patterns can be seen in those Antarctic and Greenland ice cores. It is a well-aligned rising and falling of atmospheric carbon dioxide and global temperatures as the world moves in and out of ice ages. The tight correlation holds even in minor climate swings.

But this alone doesn't prove that more carbon dioxide equals a hotter planet, something that climate scientists know well. Instead, the connection between any greenhouse gas and the atmospheric heating they all create can be shown in basic physics experiments.

It's called the greenhouse effect because it works just like one of those hothouses. The atmosphere lets a considerable amount of shorter-wave solar radiation pass through and heat the planet's surface; that heating, in turn, causes Earth to emit longer-wave infrared radiation back to the air, where greenhouse gases trap some of that heat and send it back toward the surface. In fact, without a greenhouse effect at all, our planet would be uninhabitable—a big ball of ice. Water vapor accounts for most of the heating, which allows life to flourish, but the other gases—carbon dioxide, methane, nitrous oxide, and more—matter too. The problems start to arise when we dump too much of them into the air, too fast.

It often surprises my students to learn how long we've known this. Way back in the mid-nineteenth century, amateur scientist Eunice Foote and Irish physicist John Tyndall separately conducted simple experiments to show the connection, and both even speculated that rising concentrations of greenhouse gases would warm the planet. As is too common in science, Tyndall tends to get the credit, but Foote actually did her work three years prior, in 1856. She filled glass containers with "common air," carbon dioxide, or water vapor. Then she exposed them to the sun. The latter two containers heated up far more than the common air cylinder. We know now that was not because of sunlight directly but because of how the infrared radiation that would also have been present interacts with the bond structure of greenhouse gas molecules.

So where's the correlation versus causation trap? It rears its head in one of the frequent arguments made by climate change deniers. If you look at the rise and fall of both temperature and carbon dioxide in those ice cores, temperature almost always leads the way. See, they say? Temperature *causes* carbon dioxide to go up, not the reverse! And it gets tricky because they are partially right: through a series of feedbacks that occur in both the ocean and on land, the warmer it gets, the more CO_2 tends to end up in the air. But here's the critical difference: thanks to Foote, Tyndall, and many more scientists since, we know that CO_2 itself *does* cause atmospheric heating, creating another feedback loop that raises the temperature in and of itself, accelerating the whole process.

Sometimes we fall into the cause-and-effect traps by mistake. Sometimes those traps are deliberately exploited, as with the culture wars surrounding climate science. And sometimes the trap can break our hearts. Like it did Neva's and then mine.

After the seizure in her office, Diana went downhill quickly. With each passing day, it became harder and harder for her to make the words in her brain come out, be that through keyboard or speech. The latter was now halting, each word a visible effort.

At one point she fell down a portion of the staircase that led from our bedroom to the main level. She ignored my pleas to call when she wanted to come down, so I installed a second railing. As it was on her failing right side, it was of little use on the way down, but at least it helped her climb back up and gave me something tangible to do. Something to create a momentary distraction from my feelings of powerlessness.

Soon after her fall, my father bought a one-way ticket from Montana and moved into our guest bedroom, there to help however he could. One afternoon I walked outside to find Neva and her seventy-five-year-old grandfather jumping ceaselessly on the trampoline, with Diana sitting nearby. Neva bounced off her butt and then back onto her feet before turning toward her mother.

Mama! Come jump with us!

I started to intervene, but Diana held up her good hand, then slowly stood up from her chair.

Okay. Love. Help me . . . up . . . the ladder?

Neva reached down and grabbed Diana's left hand as my wife struggled through the small gap in the netting and managed to get to her feet. Then she willed herself to jump with our daughter, her right arm largely useless for balance and her left holding one of Neva's, each leap slow and low and on the edge of a fall that somehow never came.

Every morning Diana and my father would take a walk on the East Campus path. Diana would be determined to make the circuit, pushing things to the edge just as she had right after her surgery. I could scarcely stand to watch as she swung her right leg stiffly in robotic arcs, her foot often dragging along the gravel. My father would hold her left arm patiently, taking each step beside her, ensuring that she did not trip.

When they'd return, she would sit in a brown leather chair in the living room, sometimes struggling with the iPad communication exercises, sometimes just staring at the fireplace with distant eyes. Occasionally she would try to read, but she found it difficult, and she had no patience for watching movies or TV, seeming to regard them as a waste of time that was now far too precious.

For the first time since I'd known her, that endless curiosity appeared to be slipping away. She came across as depressed, which was understandable but also terrifying. I kept trying to find ways to break through, but her growing inability to communicate made it hard to know what—if anything—was finding purchase. And it was all made worse by Neva beginning to shy away. The combination of a partial paralysis on the right side of Diana's face along with her speech deficits scared our daughter, which was also understandable, but it was heartbreaking to watch.

I racked my brain for things I could do to help them both, and learned that reading aloud seemed to help a little. I'd show Diana papers from her desk or books from our shelves and let her pick something. Half the time she would fall asleep as I read, but I took that as a sign that she was able to relax and find some peace. The best of the reading sessions came from Neva.

I'd have her select a book and then sit on a chair in front of both Diana and me while she read it to us. These moments would elicit a lopsided smile from my wife and proved enough of a distraction for Neva that it let her set her fears aside in favor of pride in her reading skills.

Still, I could scarcely begin to imagine what was going through the head of a six-year-old girl who had endured the trauma of her own brain tumor and now was watching one turn her mother into something fearful. How could she not have an absolute cauldron of emotions going on within, the full nature of which I'd probably never know?

As for Diana, over the past nine months she had risen above her likely fate with a level of grace and courage that left many simply astonished. She'd received dozens of notes from people after the Angels Among Us race, and it was becoming clear she was a growing inspiration to people we didn't even know. But now the fuel that powered it all—her scientific curiosity and the ability to feed it almost whenever she wanted—was being cruelly stripped away.

Amid these purgatorial days I tried to find some purchase at work. And I largely failed. I sat in meeting after meeting and remembered almost none of what was said as soon as each one finished. I'd overreact to a routine problem. I'd sit and stare at a document on my monitor, unable to do anything with it, before sweeping it aside to do yet one more search on immunotherapy-driven swelling in the brain. Mining the literature for even the smallest nugget of hope. Eventually I'd give up and walk the campus, drifting from path to path. On one such walk I ended up in Duke Gardens again, and jumbled memories of the surgery day came flooding back. On another I found myself in front of the biology building, where Diana's office was still untouched after the seizure. I sat on an anodized bench and wept.

Her vaccine trial was designed to be three successive shots, each about a month apart. The final one came in these late-October days. She lay back at a slight angle and breathed through the pain of its injection. Then it was around the corner to another room, where we waited for Henry and Minh. They swept in shortly, Henry as always starting with a hug for each of us against his ever-present white hoodie. He pulled up a chair and began to talk, his words more measured than before.

Look. This is not going as we hoped yet, but it's not over. We need to see what this last shot does. Okay?

Diana just nodded. I asked, not for the first time, *Could all this be an immune reaction?*

I was scrambling for a tenuous foothold in the land of cause and effect. I knew what the vaccine was supposed to do. And I knew that Diana was now having a bunch of worrisome symptoms after those vaccines began. I wanted to believe this particular correlation had a causal effect that would diminish our worry and even provide evidence that the whole thing was working. But there was no way to prove it, and Henry put a bit of a damper on my hopes.

Maybe. It's not out of the realm of possibility. I'll be honest—if that's what it is, it's lasting longer than we would expect. But it still could be that. Partly why we need to watch this last shot carefully and then see what the next MRI tells us.

When will you do that scan?

Probably end of November, more or less. Give things about five weeks.

He looked at Diana.

How are you feeling?

Okay. Well. Not. Really. This sucks.

Yeah, I know. Are you doing the therapy?

No.

Why not?

I. Hate it.

Understood. You've earned the right to make your own choices. Truth is at this point it might help you some, or it might not. Like I said at the beginning of all this, you need to make the choices that are best for you.

Can I. Choose. Not to. Have. This. Fucking thing?

She smiled her unbalanced grin.

Henry laughed.

We're working on that.

Then he turned serious again.

We are not out of options yet, okay?

Time with Henry always left us feeling buoyed. That evening we sat with Neva in her bedroom as I glued Wonder

Woman decals onto a pair of bright red rain boots, the final piece of her Halloween costume. We watched her don the full outfit and then stomp about theatrically.

Dada, how many more days until Halloween?

Three, honey.

And everyone comes to our street, right?

The one right behind us. Just down the alley.

Are you coming with me?

Of course.

Mama, can you come?

Diana smiled.

I will. Try. Honey.

But when the day came, she couldn't go. She posed beside Neva for a photo, gave her a big kiss, and told her she couldn't wait to hear all about it. That all was okay and she just didn't want to slow us down. Only I knew how much it was killing her to stay behind. How much effort was beneath the facade. When Neva returned and rushed upstairs to show Diana her candy haul, she found her angled across our bed, uncovered but sound asleep.

Dada, is Mama okay?

She's just really tired, honey. Her medicine makes her very tired.

My daughter, wise beyond her years, looked skeptical, and the questions continued as she brushed her teeth and readied for bed.

When is Mama going to get better?

Why does her smile look funny?

Why can't she walk right?

The questions carried over to her bed, where I lay beside her for an hour or more, trying to answer as best I could. I talked to her yet again about how Diana had a tumor, how they can do things to our brains that change our bodies.

Is it like mine, Dada?

Well, sort of. Not the same but sort of.

A long silence and then she crushed me with, *Dada, did Mama catch her tumor from me?*

I held her tight and whispered, *No, that's not how it works,* again and again, until she finally relaxed and drifted off to sleep.

There it was. Neva's evolutionary wiring had led her into the cause-and-effect trap, burdening her with the most gutting of conclusions. Because, especially in the mind of a child, the world is nothing but causation and pattern. Science, though, tells us—hell, it tries to hammer home—that correlation is *not* causation. Which in turn lets us see that two souls can be correlative in life, can suffer alongside each other, without one having caused the suffering of the other. In this, we find both the limits of science—we can't control everything—and also the mercy: we don't cause everything.

CHAPTER SEVENTEEN

At the end of Saint Paul's most famous passage, he says: And now these three remain: faith, hope, and love. And the greatest, he says, is love. Thinking about how the elements we all contain are but temporary combinations and by that very fact connect us to the past and future helps me find my own comfort from his words. Because love is the only part of us that can truly enter deep time. When our faith and hope and knowledge are gone—even when our brains and bodies are gone too—our love remains, moving through the world.

Scientist and poet Rebecca Elson put it this way, as part of her stunning poem "Antidotes to Fear of Death":

No outer space, just space,
The light of all the not yet stars
Drifting like a bright mist,
And all of us, and everything

Already there
But unconstrained by form.

We are all bound by the inevitability of someday confronting our mortality. Some seem to think that legacy is a way to cheat death, that if the things we accomplish or accumulate are notable enough, it's a ticket to a type of immortality. But from an atomic perspective, our immortality is already guaranteed. When the roughly seven billion billion billion atoms that each of us contains are assembled into our bodies, they bring with them a far greater number of stories from the past, add ours during their short stay, and then go on to accumulate an unfathomable number more. And yet each one of those stopovers is unique. As Carl Sagan put it:

> Every one of us is, in the cosmic perspective, precious. If a human disagrees with you, let him live. In a hundred billion galaxies, you will not find another.

Perhaps that's part of the reason we grieve so deeply when those we love pass away. We know that right down to their smallest building blocks, they are literally irreplaceable. This is no doubt why religions that center concepts of heaven or rebirth have so much appeal and why these themes exist across all major faith traditions. It's a way to circumvent the basic rules of life on Earth, a framework meant to soothe us with a belief that there are ways for our loved ones—and ourselves—to carry on.

In *Horizon*, Barry Lopez writes that *the world outside the self is indifferent to the fate of the self.* Yet our selves are inextricably, repeatedly, wondrously connected to that world. I find comfort in our inevitable disintegration and sometimes think of us as a seed source of nearly infinite potential. Because my atoms and yours and everyone else's will go on to play a part in every story the world has yet to write. That, for me, is the permanence of our love.

I began to tell people they should visit soon. That we had not given up hope but that Diana was also declining into a state where communication was increasingly difficult. That this might be their last chance to have a conversation with her. Each call was as gutting as those when we first shared her diagnosis.

They began to come. Her sister and niece, Elene and Sofia. Her mom and brother. Friends from Boulder and my two older kids, Kaelan and Lily, both of them years older than Neva but deeply bonded to her. Even my ex-wife, Sue, came, reminding me once more how cancer can surprise you in all kinds of ways. My relationship with Sue had been largely civil since our divorce, but it was complicated, too, and mostly consisted of me trying to navigate two separate worlds. Sue softened notably when Neva fell ill and had always treated her with love, but I didn't expect her to be a pillar of support when Diana's diagnosis came. I was wrong.

The adults would sit with Diana and talk or just hold her hand or usher Neva to moments of distracting fun. By Thanksgiving, the conversations were almost entirely one sided. The right side of Diana's face was now in near-full paralysis, and even her comprehension of what others said seemed to drift.

And so, when that first Tuesday of December arrived and Henry walked into the room, I already knew he would tell us the tumors were far worse. He sat and held Diana's hands, and he told her there were still things they could do. But that, barring a miracle, none of them could cure her. That all of them would increase her need to be in the hospital and bring new discomfort and pain. The price of trying to extend a life.

She sat and looked at him through welled eyes and let out a single word:

Neva.

Then she looked at me, pleading for help. Wanting to say more but unable. I swallowed hard and struggled to keep my own composure as I asked Henry to elaborate.

What are her full options?

He seemed to understand and went to the heart of it.

Pretty much everyone with one of these tumors hits this point. You need to make a decision about quality or length of life. We can probably buy you some more time. How much, I don't know. Maybe weeks. Maybe months. But it won't be easy. Or you can decide to stop treatment. If you do that, you can enter hospice care at home, and our goal is to keep you there as long as possible. Hopefully for the entire time.

He paused, his eyes laden with an emotion we had not yet seen.

Why don't I give you a little time to think about it?

Diana reached out, grabbed his arm, and let out one more word:

No.

I'm sorry, I don't understand. No what?

No. More. Go home. For. Neva.

Somehow I drove us there knowing my wife would soon die, feeling the full weight of it finally break through the last shreds of my hope and denial. But as we pulled into our driveway and I went to help Diana out of the car, I saw a newfound peace on her face.

More calls, more tears, more visitors. Some came and went, struggling to keep things bright as they said goodbye. A few key friends made it clear they had no intention of leaving, joining my father. My mom, who endured the pain of being apart from all of us while tied to her work, arrived as the holidays approached. The house was mercifully full and a little chaotic—something that we—that I—desperately needed.

At first Diana maintained her daily trips downstairs to sit in the big leather chair before the fire. But her stays grew shorter until they ultimately ceased. She remained upstairs, eating less each day, visitors cycling through the bedroom to sit quietly at her side or perhaps tell her a story or two. At some point a book got made; I cannot remember how through the searing and sometimes beautiful images of those final weeks. The book had a simple white cover that belied the depth of what was inside. Picture after picture of Diana and Neva, or at times the three of us, spanning the years from the day Neva was born. The two of them in matching Mardi Gras dresses, making faces at me across a DC restaurant table, or laughing together side by side on a pair of swings. Neva brought her mother the book, and they lay beside each other, turning the pages, her little girl overcoming her own fears to tell Mama about each picture. Diana kept the book beneath her pillow.

When she became unable to walk to the bathroom safely, we set up a portable toilet beside the bed. We would lift her carefully and set her down while she slumped to her left side against the handrail of the commode. Then back to the bed. Soon after she became incontinent, and a new routine began. Large absorbent pads adhered to the bed, one of us rolling Diana to the side and cleaning her while the other changed the pad out. Her body became increasingly skeletal. By Christmas, her hip bones were shockingly pronounced, and each of her ribs was clearly outlined beneath papered skin. And yet she sat up and smiled as Neva bravely insisted on bringing all the gifts from under the tree to the bed. As she sat chattering and helping her mother tear the gift wrap from the picture book they'd already shared.

Mama, I know you like this book, so I thought you might want to open it on Christmas.

That night Diana took my hand and struggled to say two names.

Lily. Kaelan.

They are coming, honey. This week.

She closed her eyes and seemed to fall asleep. Then she opened them, looked pained, and tried to speak again.

Mor…

More what, honey?

A slight shake of her head.

Morph…

You want the morphine?

A nod. For the first time, I pulled out the vial of pale blue liquid and drew a small aliquot into a syringe. My hands were shaking. Then I slid it between her lips and depressed the plunger. The lines on her face faded, and her eyes closed once more. Her breathing became so quiet that I panicked, momentarily believing the medicine had ended her life. But then I felt a soft breath on my cheek and lay back beside her, eyes open for hours, straining to feel the nearly imperceptible rhythms of her sleep.

By the end of December, the last of the visitors had arrived, and on the afternoon of December 30, the hospice nurse came. As before, she replaced a few supplies and took Diana's vital signs. I told her Diana had essentially not eaten for more than a week and that I felt like the end was very near. She told me not to be so sure.

I know it seems that way, but her vitals are still strong. I've seen patients like this hold on for weeks, sometimes even start eating more again. It's hard to predict, but if I had to, I'd say she still has quite a while to go.

Yet as I sat beside her the next morning, soon after bringing Neva in for another visit, I knew something had changed. Her face darkened and her breathing became more labored. I called the nurse and told her I thought she should come right away. By the time she arrived, Diana's breaths were even more pronounced, her skin tone darkening further. The nurse quickly assessed her vitals once more and verified the change.

I'm really surprised.

I wasn't. I realized that this astonishing woman was saying goodbye on her own terms. She waited until virtually all those who mattered in her life could visit, she held our daughter a final time, and then somehow willed herself to spare all of us the prolonged pain of an extended goodbye. In ease and acceptance, she found incalculable power.

How did she do it? I can never know for sure. But I'm convinced that her devotion to science as more than just a source of proximate answers but as a way of being in the world—and as a way of being with herself—allowed her to become something extraordinary when it mattered most. Her daily practice of approaching life through a scientific lens helped her to find peace and courage in ways that profoundly shaped not only her final days but the paths of the people she loved long after she was gone.

Even in this moment of absolute anguish and loss, her evident peace gave me something that I could then return. I sat beside her and held her hand and stroked her forehead and told her again and again and again that she was free and that we would be okay, that she was forever burned into every last inch of our souls. She drew a final deep and shuddering breath. I reached up and gently closed her mouth and eyes.

I sat and looked at her lying there tranquilly for a very long time. Then I turned to face the hospice nurse, who had stood quietly in the corner throughout Diana's final moments. There were tears on her face. She said, *I've been doing this for years. But I don't think I've ever seen someone die so beautifully and well.*

A few weeks before she died, Diana asked me to write this book. At the time, she said:

Please find a way to make our story help someone else.

I remember looking back at her in wonder and thinking about how many times I'd watched her lift someone else up. How it came to her as easily as drawing a breath. And as the years have passed since, I've come to believe that living and dying as she did are not impossible, are not reserved for the almost mythically strong among us. Because in truth, while she will always be deified in my own mind for what she brought to my life, she was just a person, like any of the rest of us, but perhaps differentiated in her ability to use one of humanity's most wondrous constructs—science—as a path to developing her extraordinary soul.

CHAPTER EIGHTEEN

On a small island not far from my Montana home, there is a forest in apparent ruins. Like so much of the West of late, a fire swept across the island in August of 2020, leaving behind a dystopian landscape of blackened trunks and drifting airborne eddies of fine gray ash. In some places the fire burned so hot that trees vaporized into the sky above, leaving holes in the earth where even their roots were consumed. When I visited the island in late September of that year, my boots disappeared into the ash below as I climbed through the charred remains.

And yet there are pinecones buried in that ash. As it turns out, some of those cones will only open to release the seeds they contain when fire burns their parents to a crisp. In the language of ecology, the cones are known as serotinous, a word whose literal meaning is "remaining closed with seed dissemination delayed." Within a few years, I will be able to go back to that island and find a thousand needled seedlings, each one jauntily reaching for the sky.

Many of us like to believe that when the worst hits, we can simply power through. We don't want to think about how our lives may hang in the balance. We don't want to admit that a true rebirth may not occur until we are nearly burned to the ground.

I mean, how do you tell a little girl her mother has just died? I remember Neva's mixture of confusion and acceptance, the latter especially striking, the wisdom beyond her years once again showing up. But I don't remember much else until the moment the doorbell rang, and I left Neva with Greg's wife, Robin, and my dad to keep her from seeing the two somber men who ascended our stairs.

Not long after Diana's final breath, Robin and I had pulled one of her favorite dresses from her closet, the teal cloth hanging loosely on her wraithlike form. Now I watched as the men lifted her from the bed and placed her on a foldable stretcher of nearly the same shade as the dress, a stark contrast with her dark hair. The top of her head rose in and out of view as they struggled to take her down the stairs while I trailed behind, as though I had some useful role to play. Knowing I did not. I followed them out to the alley anyway, and we all stood in a steady and cold rain as they slid her into the back of the hearse. The driver, a stocky man with sandy hair, took my hand.

We'll take good care of her.

Then she was gone.

I stood for a while amid the spiderweb cracks and filling puddles of the alley beside our home, letting the rain soak through and welcoming the creeping numbness. When I walked back inside, I smiled wanly at loved ones who were uncertain what to do or say, then pushed through the sudden hollowness of our bedroom and slid into the shower. I stayed there until I could muster enough resolve to find my daughter once more and try to take the first real steps in a life I still could not imagine.

By the next morning, Diana's request for the Call and Gus protocol gave me a fragile anchor. I spoke with the mortuary about how she could be flown to Boulder; our friend Scott somehow found a New Orleans second line band. Another friend had already launched the idea of silver bracelets engraved with an outline of Mount Neva that my daughter and I drew together, and by the time we arrived in Colorado, they were ready. One for my wrist, one for Diana's, and a smaller one for Neva.

Six months before, Diana had given me guitar lessons for Father's Day. By early fall I managed to write her a song, playing it for her one night on our screened porch. A hum of cicadas as a backdrop. She closed her eyes as it finished and then asked me to play it again. When I did, she said, *If the day comes, I want you to play this at my funeral.*

At the time her request made me shelve the song in despair and denial, but now I practiced and practiced and wondered if I could pull it off. Then I pulled in Scott and my former PhD student Phil, both far more accomplished musicians than me, as an accompaniment.

The day arrived cold and gray but windless, and as the swelling crowd assembled outside the cemetery office, light snow began to fall. The band led us slowly up the short hill with a piercing rendition of "A Closer Walk to Thee," the notes ringing through the cold air with a resonance I could feel. Each of my three children marched slowly by my side. When we reached the upper-right corner of the cemetery, the one that gave way to her test run, I broke off with Kaelan and Lily and a few of Diana's closest friends. We lifted her casket from the waiting hearse and carried her to a platform above the newly dug grave.

Greg stood on the slope above as the crowd encircled her on the other three sides, and he began to speak of life and love and the ways the woman before us would carry on. And then my two older children each stood and recited tributes they wrote themselves, both of them sending their words through and above and around us with a power so simultaneously devastating and uplifting that when it became my turn, I didn't know if I could even utter a thing.

But the words somehow flowed, and so did the music when I sat between Scott and Phil, all of us playing her song as she was slowly lowered to her place of rest. Only when the song finished and she descended from sight did I begin to disintegrate.

The band led our march back down to an instrumental of "Do Whatcha Wanna," at which point a few smiles began to appear, and by the time we reached the parking lot, musicians and mourners alike were dancing in the snow to a rollicking version of "When the Saints Go Marching In." I stood to the side, taking it in, knowing she would have wanted it no other way.

Years before, I sat with Diana on a sage-covered slope in the northeast corner of Yellowstone Park, watching wolves. For a while they and we were equally immobile, all of our breaths suspended visibly in the quiet of a subzero December day. We were cold, but the pack—eleven of them all told—was at ease. They formed a monochromatic arc across a snowbound hillslope, each black, gray, or white dot brought into stunning detail by the spotting scope before us. Occasionally, one would arch a back or scratch an ear, but that was it.

Until a coyote showed up.

The intruder picked his way along a swale far below the pack, occasionally cocking his head and thrusting a muzzle into the snow, while twenty-two eyes tracked his every move. Eventually the insult became too great. Three of the wolves began a steady and purposeful trot down the hill. The coyote stood tall, marked their approach...and ran like hell. Apparently convinced his prospects were dim in the open riverlands and blocked by the wolves on the valley's southern slope, he sprinted straight for us, veering late to cross the road in a mad dash for the lodgepole pines above. He made it but with the wolves close behind. When the three wolves returned to the valley a few minutes later, there was no sign of the coyote.

It was a stark reminder of how their world had changed. Soon after the reintroduction of wolves to Yellowstone, elk and deer and coyotes who once did this...now had to do that. In turn, many plants and animals that for so long just couldn't... well, now they could. Add it all up, and within only a few years the whole place was different. In the language of ecology, Yellowstone wolves are now an oft-cited example of a keystone species. When they are there, you get one thing. Take them out, and pretty much everything is different.

I thought of Yellowstone as I stood in Diana's home office back in Durham after the funeral.

Now what?

We were there and she was not, and we had to carry on under an entirely different set of rules. Not only that, but to meet one of her insistent requests, someday I had to start a new life I never foresaw or wanted.

Then I saw one of her running medals on the desk and thought, *Maybe I'll start there.*

Never the runner she was, I nonetheless set out on a maze of forested trails just across the Durham County line. Slowly at first and then with increasing sureness, a planned three-mile loop becoming five, then seven, then a turn away from the waiting car to do the five-mile loop yet again. We had run these trails together during our first fall in North Carolina; now they were a way to keep her beside me.

Back home, the run's buoyancy let me begin the work I dreaded. I boxed up some of her clothes to keep for Neva, separated others to donate. I cleaned out her desk. I even managed to laugh when I found a hidden bottle of THC-blended gummy fish that two of her former students smuggled out of Colorado, just in case it helped. At the time, she relayed the story of her chemo-medicine-induced descent into paranoia and told the students she had to steer clear. But the bottle contained only a single remaining fish.

The routine carried on day after day, run then sort, keep moving forward. One day I gathered and disposed of a box full of her old medicines, starting by pouring the morphine down the drain even though I knew I should probably find another way. On another, I launched the probate process and started the paperwork to transfer her retirement funds. Daily reminders of what no longer was, softened just enough by the ways each run still kept me tethered to her.

By the end of January, I was running out of obvious things to do, and on the one-month anniversary of her death, I found myself drifting down hard. My thoughts wandered, stuck for long periods in abject despair, but, incongruously, they landed on lessons about Earth's climate. Specifically, that a seemingly little thing can lead our planet down a very different path, for a very long time. A certain slice of the ocean gets just a bit less salty. A reservoir of frozen gas, trapped for eons, begins to leak. Or one of my favorites: this lovely and always askew home of ours tilts just a little bit less on its axis.

Do yourself a favor and google "Richard Alley congressional testimony," and you'll see the best and most entertaining explanation of this phenomenon out there. In less than thirty seconds, the famed glaciologist responds to some dismissive questioning from a congressman by positioning the hostile representative as the sun, while he uses his own bald spot as the North Pole, nodding his head to and fro as he explains how the planet tilts back and forth as part of something known as the Milankovitch cycles.

That tilt is not a sudden event. These cycles, named after the Serbian astronomer who first described them way back in the 1920s, take far longer than any one of us will ever witness. But tilt back and forth we do. Over thousands of years, sometimes Earth stands just a little straighter up, sometimes just a bit closer to lying down. And that can mean everything.

As it turns out, the straighter we stand, the colder we get. It all starts because whenever the planet decides to improve its posture, the cold parts of Earth get even less of the sun's warmth. That means more snow and ice, which in turn bounces more heat back to space. Next thing you know, we're in an ice age, everyone standing around asking, *How the hell did that happen?*

I sat on the screened porch, drink in hand, and wondered if I would ever stand straight again. Versions of this question had orbited around my brain for a year now. They were almost unbearable in the early days, but then, amid the denial and distraction and hope, they had faded just a bit into an ever-present but more manageable ache.

Especially the hope. You see, I really did think she'd be the one to beat this. I wasn't alone. She was too tough, too healthy, too determined, too everything to not even hit glioblastoma's median survival time. But as much of an outlier as she was, so was her tumor. Of course it was. In some terribly beautiful way, because she never did a single thing half-assed. If she was going to get a tumor, she was going to get a barn burner. Henry told me as much just days before.

As aggressive as we've seen. Not sure why. Just bad luck.

So here I am, I thought, thirty-one days into a disembodied existence. Maintaining a patina of my former self and somehow navigating moments of things entirely new: powers of attorney, selecting coffins, picking up death certificates. Through it all, feeling as though I were floating above, watching a robotic version of myself go through its daily programming.

But here's the thing about Earth. After a few too many centuries in one of these tilt-induced ice ages, it too decides *to hell with this* and starts to lie back down. As though it just can't take it anymore. When it does, the ice begins to melt away. Not overnight and not without a slow and uneven climb back out. Still, the day comes when countless miles of ground once frozen and dark are ablaze in spring flowers.

The images of Earth's oscillations soothed me. *Maybe it's okay to lie down for a bit*, I thought. *Because when I do, hopefully I'll know I'm already starting to stand back up.*

Except I didn't lie down. I fell. Admittedly, I was good at rolling out a hundred versions of *no, really, I'm okay*, delivered with increasing conviction as winter bled into spring and I became more adept at fooling myself as well as those around me. But in private I suffered and slipped. Never much of a drinker, I dented bottles of bourbon each night after getting Neva to bed. All it did was make me feel worse and unable to sleep, so I piled

Unisom on top of the bourbon, stopping only when Neva shook me out of a persistent torpor one morning.

Dada, are you okay?

Mundane things took on outsized importance. In one such episode, Neva returned from school to tell me that every kid got assigned a letter of the alphabet, and on their day they had to bring in something to share with the class that started with that letter. Hers was L. I became fixated on getting her the best goddamn L choice in history. I had a dictionary on the kitchen table and a growing list of options on a legal pad.

Neva looked at me quizzically and said, *Dada, I just want to take my lion.*

It was the giant stuffed lion from her ICU days, the one still bigger than she was. She'd need help getting it to the classroom. On the drive to school that morning, she gave me a very serious overview of the class situation.

There is a Time. Out. Corner. If you are bad in class, you have to go sit in the time-out corner.

Ah, I see, hon. Have you been in the time-out corner?

Dada! NO!

Well, okay. Who does go to the time-out corner?

Mostly Luke.

Is this the kissing bandit kid you told me about?

DADA! GROSS!

Well, is it?

Yeah.

We walked down the school hallway with me shouldering the big lion until we reached her classroom door.

I want to carry it in. Emphasis on the *I.*

And so she did, to the *ooh*s and *ahh*s of the other kids, all of them seated in a semicircle on the carpet because we were running late. She managed to plop the lion onto the center table. A sandy-haired boy immediately jumped up, ran to the lion, and began punching it furiously in the face. Then, before the horrified teacher could reach him, he slunk over to a chair in the far corner, hanging his head. Luke.

I began to feel as though I needed my own time-out chair. I drifted at work, sometimes lashed out without cause, found it nearly impossible to give a damn. One day I sat and simply stared across my office, wondering if punching a lion would do me good. By late spring, I became convinced that we needed to get out of North Carolina and away from all its eviscerating memories. It was time to move us back to Colorado.

At first the return to Boulder brought a slight lift. I reconnected with old friends, rode my bike on familiar trails, and grasped a few moments of peace by standing in a river with a fly rod. And I found a home that let me run from its front door along some of Diana's favorite trails, to where I could visit her grave. But soon enough, the tide turned once more, and I began to slide even further into bouts of almost paralyzing grief. Somewhere in the middle of this time period, I read *The Dog Stars* by Peter Heller and became fixated on a line that linked my own field of biogeochemistry to a portrayal of grief as permanent. He writes:

> Grief is an element. It has its own cycle like the carbon cycle, the nitrogen. It never diminishes not ever. It passes in and out of everything.

The bottom came in the form of a nondescript and cedar-planked storefront, unknown to me before a brief and gutting online search. I sat across the street in my truck and stared vacantly at its drawn windows. Eventually I opened the door and crossed slowly. I stood on the doorstep for several seconds before entering to a dimly lit world of racks and shelves that bordered a wood-framed glass case. A display of glasses rested on one end of the case, the lenses yellow and orange and pale brown. The wares on the walls were garish and large, unfathomable, but those within the case felt more accessible. I stared at them, vaguely aware of a ponderous man behind one corner of the glass, a graying ponytail and beard below a weather-beaten ball cap.

Can I help you with sumthin'?

Um...yeah. Can I see this one?

The gun was slate gray and heavy. I held it like it might strike me and wondered how it would go. I remember feeling a crushing weight inside and a fogginess in my mind that left me unable to do anything but just stand there and stare at what was in my hand. When the man spoke, it seemed to come from very far away.

Ya want me to start the paperwork?

And with that, my scientific brain returned. I was back on familiar ground, questioning my conclusions. Suddenly I could step beyond myself to see the trajectory of my descent and the horrific pain it would cause if I couldn't stop. I dropped the pistol on the glass counter and stumbled for the door, nearly falling down the steps before running across the street and collapsing into the cab of my truck. That night I lay awake beside Neva and held her as tightly as she would stand, my arms in a welcome ache by the time dawn arrived.

CHAPTER NINETEEN

If you know the story of the American chestnut at all, you probably know it as one of tragedy and failure. This magnificent tree was an icon of America, such a dominant part of Eastern forests that to think of them without these trees would have been, well, unthinkable. Then came the chestnut blight on top of all the other assaults we brought to the woods, and poof, like that, they were nearly gone.

And yet, once again, a blend of science and love has introduced a new twist to the story. With modern genetic tools, we can slot in a few genes to the chestnut they never had and never would, and within those genes lies the potential to survive the blight that has taken them down. Meanwhile, the chestnuts are there still, rising to die, doing just enough to have a few new seedlings emerge. The serrated edges of their leaves seem appropriate, as though each one is defying the rules of physics to gain purchase in the very air that both allows their existence and carries their fate. Somehow, they barely hold on, waiting for that day when they are transformed right down at the molecular level and, through that change, are able to come home again.

In this the chestnut story is a microcosm for science as a whole: ultimately, its purpose is to bring just a bit of the chaos together where we can and while we can, so that the possibilities of the world multiply. Those new genetic discoveries are creating prospects for the chestnut that previously did not exist. And isn't that what we really want—a world of possibility? A world where our stories are neither predictable nor foretold, but one in which our own actions and pursuit of knowledge can open up paths we didn't know were there? For me, this is science at its most beautiful and fundamental: a means by which we expand and enrich the potential canvas of our lives.

Walking up to the edge of ending my life changed me at a molecular level too. This darkest hour opened the door to the beginnings of a peace I'd never known. Like the very elements that construct us, I was starting to see how we can emerge from the ashes of the worst moments in our lives as a wholly unexpected fusion of past and future.

In *The Solace of Open Spaces*, Gretel Ehrlich writes of hard-won comfort after the death of a loved one and of how her eventual peace is forged from the absolute indifference of a stark yet beautiful Wyoming landscape. Paradoxically, given the title, she also writes that *true solace is finding none, which is to say, it is everywhere.*

For me, I have also learned that solace is not independent of place. For as long as I can remember, my peace and joy are most assured beneath the biggest of skies. Psychology (pop and otherwise) likes to tell us that we are happiest when we can simply be ourselves. And it is here that another Ehrlich quote resonates most:

> Everything in nature invites us constantly to be what we are. We are often like rivers: careless and forceful, timid and dangerous, lucid and muddied, eddying, gleaming, still.

For the most part, I flow best and clearest either when crisis is right before me or when I am in the landscapes that have always burnished my rougher edges. And so I began to run Diana's favorite trail almost daily, through the scents of pine and sage and beneath russet slabs sent aloft nearly three hundred million years ago. Along this trail, and in the town below, she was everywhere, so it was fitting that she lay right where they merge.

On most of my visits to her, I found myself saying a few words, maybe shedding a few tears, mostly just sitting by her place of rest. But a day finally came when I talked at length. I talked to her about life and love and of how she lived on in the hearts and minds and actions of so many. I told her stories of things friends and family had done since she passed that would fill her with pride. I told her about the recent horrors in the world, too, of the hate and fear and violence, of how now more than ever the world needed people who live as she did.

I started to ask her the big questions that lay before me. And as I talked of my hopes and dreams and sorrows in a new reality I still could not always grasp, I heard her once again. Heard her remind me to live boldly, generously, fully. Remind me to pour my energies into what mattered most and to lift myself by lifting others first. That there is no greater solace and joy than that which comes by being generous, by being endlessly curious, and by never being afraid to fail. To listen to what's deep inside, to question with joy instead of dismissal, and to take the big chances before me, even if some of them crash and burn. And that all of this is best done with the kind of shit-eating grin she wore better than anyone.

I knew I couldn't match the grin. But the rest? I thought, *Maybe, maybe now, it's a map I can follow.*

I needed to reach deep for that map in December as I sat beside Todd, looking once more at black-and-white slices of my daughter's brain. Her tumor had remained stable for more than a year. No longer.

You can see the new growth here.

Yeah. Shit.

I paused, swallowed.

What do we do?

Well, I still feel as I did before. Even better now since she's grown a bit. She's an unusually good candidate for a second surgery, and if you look at the remaining mass—he pointed at the screen—*it's not big and it's in a place where I think we have a very good shot at getting all of it this time. I can't promise that, of course. But the chances are good.*

I sat and let that sink in. The moment of telling Neva she needed surgery again, the terror of waiting for her to emerge. But then I began to think of her as tumor free, of what that would mean, how it would not remove the possibility of growth again but would make it far less likely. I squared up.

Okay. When?

We don't have to rush, but I wouldn't wait too long, either. I'm thinking late January.

On New Year's Eve, I sat in a snowbound cabin just west of Mount Neva. I woke before dawn on this one-year marker of Diana's passing with the mountain that is our daughter's namesake visible from my cabin window only as an outline. A thin and partial corona from the cities on its opposite side revealed the summit but did not yet obscure the brilliant splash of stars above. Soon enough, the stars faded and the mountain began to define itself, first softly and then with increasingly sharp, rose-tinted detail. It was capped with snow, but beneath lay a basic form that is around three hundred million years old. In parts of the mountain there are veins of rock that date back to the dawn of multicellular life, more than a billion years before the first real ancestors of the trees that now line its flanks.

In short, Mount Neva has seen some things. Dinosaurs in its valleys, new mountain ranges popping up around it, the rise and fall of a great inland sea. The upheaval of my own last few years would not even move its needle. But the constellations above its dark shape reminded me that the dust of long-distant stars still lies within its ancient rocks, within the snow and trees above them, within me and my daughter and the extraordinary woman who left a year before. The thought, coupled with the first rays of sun behind the peak, brought me a peace I did not expect.

As I looked at the mountain, I thought back to that California lecture hall all those years before and the moment when the stupid bumper sticker version of the stardust metaphor became something far more meaningful. How our lives are just a blink of a geologic eye, and we can still find comfort in that fact. In his book *Underland*, Robert Macfarlane writes magnificently about human existence in the context of deep time, at one point noting:

> We are part mineral beings too—our teeth are reefs, our bones are stones—and there is a geology of the body as well as of the land.

I looked at Mount Neva and thought about how for me, Diana had brought a new twist to our lives being just tiny blips in the grand arc of time. She reminded me that while our flashes may be brief, some of them are impossibly bright, and everything that matters is contained in the ways your own light sparks the ones that lie in everybody else.

A year after her death, I was seeing Diana's light still burning not only in people I knew but in others who touched her orbit. One of those who knew her, but not me, wrote to express her own grief in part by saying, *How sad, for she was destined for greatness.*

I wanted to write back with:

No, she already arrived.

Over that year, my own light flickered frequently, sometimes frighteningly so. In a year of upheavals, some of the worst moments came when my daughter, mostly built of Diana's same spirit and strength, crumpled under nightmares that I would follow her mother into the grave. But as I looked at the mountain that morning, I remembered that we each had more than a dusting of the light Diana put in us and that Neva's is unquenchable.

And I thought to myself, *This surgery is going to work.*

Just a few weeks later, I struggled to remind myself of that faith as the anesthesia made her eyes roll back into her head once more. I kissed her goodbye and walked out of the stark operating room with the lasting image of those untethered eyes. Then I steeled myself for the torture of sitting and waiting and trying not to think of her lying on that glaring antiseptic table with a hole drilled through the back of her sinus cavity and an array of cameras and cauterizers and suction devices tracing a path from the outside world to her brain.

Prior to her short trip from the prep room to the OR, she high-fived the anesthesiologist. She laughed at Robin and me dressed in our surgical bunny suits. She set her jaw as she had too often for the past three years and radiated a strength that made even the hardened veterans of a level-three hospital step back and comment.

How old are you again, sweetheart? My goodness, you are brave.

Her mother's child.

A few nights before the surgery, she woke up with a nightmare, shook it off, and began to talk matter-of-factly about how it would all go down.

Wait, Dada, so they will make a hole to get to my brain and then put a camera in there?

Yes, love.

And then how will they get the tumor out?

With special tools, almost like a tiny vacuum.

Will I have a tube in my arm when I wake up?

Yes, just like before.

Can I watch movies?

Of course, honey.

Okay, I'm going to sleep now.

I lay beside her in the dark and shook my head in wonder as I had a thousand times before. On this night, lines from an old Emily Dickinson poem came to mind.

> *Hope is the thing with feathers—*
> *That perches in the soul—*

And sings the tune without the words—
And never stops—at all

Then I slept too, somehow knowing that she would be okay because nothing else was acceptable. To her.

And she is. Before I know it, she is parading about the house with her new puppy and skiing the mountains to our west and biking the deserts of Moab beside Kaelan and Lily. Days before her eighth birthday approaches in June, she puts her face tentatively into a gently swirling patch of Hawaiian sea, the mask slightly askew, and struggles a little to breathe through the plastic tube. We walk out a few more steps, try again, and for the first time she jerks her head up in wonder and lets the snorkel fall as she says, *Dada, I can see the fish so close!*

Now we are beyond the others, the shallow reefs giving way to deeper canyons of darkening green. Thirty, forty, even fifty feet down. She is unafraid and lost in the marvels of a new world. When we return, she talks of seeing an eagle ray, and when I doubt her story because I saw no such thing, she describes the animal and setting in perfect detail. The next day we enter a stormier sea through a channel in the reef, and as I help her, I embed the wrong end of a sea urchin in my butt. Later, she laughs while I sit in a bowl of vinegar to dissolve the black spines as she describes yet another ray she saw and I did not.

On the following morning we head south, climbing past Kona as the landscape shifts from dried and exotic grasses on jumbled lava flows to the variegated greens of coffee plantations and patches of forest. A belt of native ohia trees rises above another more recent lava flow, their flowers aflame, before we can see the windmills of South Point and then the astonishing bulk of Mauna Loa behind the verdant pastures of Na'alehu. As we enter Volcanoes National Park, I think of how the elements within the leaves of the forest around us record the ups and downs of their own lives.

When times are tough, they respond by changing their molecular construction to encase themselves in a waxy hardness, a shell against the onslaughts of the outside world. They take on a hue that is nearly gray, as though reflecting a surrounding environment that has lost its full expression. You can pick one of those leaves and put it on your table, where it will remain hard and dull and nearly unchanged for months. But when the rains come and the temperature is just so and the soils improve, those same trees become flush in leaves that are a brilliant shade of green. Pick one now, and you'll find it wondrously pliable and soft.

When we arrive in Volcano, a mist hangs over the town, but it is punctuated by moments of blue. We are with my parents, and I tell them I would like to go for a run. It is the one I did with Diana when I turned forty, ten miles or more of primeval rainforest that gives way to landscapes of the moon.

At the most abrupt of those transformations are steps. They are steep and rough-hewn conglomerates of a thin layer of soil and the lava on which it rests. They exit the forest and plunge toward a dark and crevassed expanse below, where a few pioneering trees barely reach waist high but explode in red flowers all the brighter for their stark surroundings. As I reach the path that bisects the crater's base, a laden sky opens, and I am running, running, running, my arms open wide, the sheets of rain tinged with something destined. With the knowledge that she will course through my veins until my own elements are scattered to life not yet formed, but so, too, that new stardust will someday appear.

ACKNOWLEDGMENTS

This book would not be in your hands without the brilliance, patience, and enduring support of two people: my agent, Anna Sproul-Latimer, and my editor, Maddie Caldwell. Like the gnarled little tree that appears in chapter twelve, Anna took an idea that clung to an outpost of hope and pain and helped grow it into something far richer than what I first put down on paper. Then Maddie's faith in the project, and her exceptional editorial insights, transformed it once more. All of my scientific papers have coauthors and deservingly so; the scientist in me feels like Anna and Maddie should be listed on the cover of this book, not just here. Given how personal this project has been and that it represents a promise to Diana now fulfilled, my gratitude to both of them—and to the rest of the team at Grand Central Publishing who helped make the book a reality—is boundless.

Kim Nicholas introduced me to Anna, another debt that cannot be repaid. Josh Schimel, Gwen Trowbridge, Val McKenzie, and Eve Hinckley read early versions and gave me an essential boost of motivation to continue. Elizabeth Kracht took it from there, serving as a developmental editor and providing my first critical insights into the sometimes murky and daunting world of agents and commercial publishing. Mark Bryant and Steve Hayward also took time out of their busy and book-filled lives to read and comment on an earlier draft. And Christine Arena and the rest of the gang involved in producing *Let Science Speak* helped convince me this might be a story worth putting out into the world. My thanks to you all.

Anyone who has been through loss and grief knows how the community around you can transform from something that enriches your life to something that becomes vital to its very existence. The list of those who kept me and Neva aloft, and who did the same for Diana while she was still with us, is long and humbling. Burke and Karen Townsend; Greg Asner and Robin Martin; Kaelan, Lily, and Sue Woodward; Elene and Susan Nemergut; Val McKenzie and Noah Fierer; Randy and Heather Chambers and their children, Maya, Alexa, and Coby; Scott Ferrenberg and Akasha Faist; Wes and Hayley Hobson; Elizabeth Costello; Linda Berman; Phil Taylor; Cory Cleveland and Beckie Rawlinson; Kate and Anna Schimel; John Parker; Andrew Todd; Alison Hastings; Laura Turcotte; Eve Hinckley; Sasha Reed, Josh Tewksbury, and Kristen Rowell; Paige Mackey Murray; Chris Gergen; Emily Graham; Joey Knelman; Dez Stone Menendez; Emily Bernhardt; Peter Vitousek and Pam Matson; Faith Cohen . . . the stardust in all of you is anything but ordinary, and you were generous enough to share it when we were most in need. And without a doubt, this list is incomplete.

Children's Hospital Colorado and the Tisch Brain Tumor Center at Duke University represent care at its finest, for in each of these places we found people whose astonishing technical skills and medical insights were matched by the deepest of humanity. Todd Hankinson, Christina Chambers, Erin Kissell, Henry Friedman, Allan Friedman, Minh Nguyen, Nancy Andrews, and so, so many more brought hope and comfort and even the occasional miracle to some of the hardest moments in our lives. And the pediatric ICU nurses we met during Neva's two surgeries were among the most competent, patient, and kind people I've ever encountered.

Colleges and universities are complex beasts that can sometimes feel too bureaucratic and impersonal. A few leaders of such institutions made them anything but. I am deeply grateful to Sally Kornbluth, Dick Brodhead, Jim White, Russ Moore, Terri Fiez, Jill Tiefenthaler, Seth Bodnar, Reed Humphrey, Adrea Lawrence, Libby Metcalf, and Paul Lukacs for all they have done to support my family and this project. The stories in this book and its subsequent writing span time at four institutions: Duke University, the University of Colorado, Colorado College, and the University of Montana. Every one of those places gave my family a home and me the space to care for them in times of need and to work on this book. The people above only begin to cover what a complete thank-you list should look like.

Science in my own life has been inextricable from the people who pursue it, and I chose the field I'm in partly because it does not lend itself well to solitary pursuits. I've always fed off the energy and company of other scientists. Once more, a thorough list is longer than I can write, but a few people stand out for what they have brought to my journey. It starts with Diana, who as I hope this story conveyed, transformed science into something of seemingly infinite spiritual depth. Peter Vitousek, Pam Matson, Paul Ewald, Dave Schimel, Beth Holland, Jane Lubchenco, and Bob Howarth mentored and inspired me, kept me going in darker times, and I would not be where I am today without them. The same goes for some of my longtime partners in research: Cory Cleveland, Greg Asner, Will Wieder, Sasha Reed, and Phil Taylor don't take it all so seriously that its practice strips away the humanity and fun, and in that very approach, their wisdom and creativity shine through and their science reaches uncommon heights.

Last and far from least, Sue and Charlie Bell became that new stardust in my and Neva's lives. They have held our pasts with grace and love as they open the doors to a future we once could not see.

ABOUT THE AUTHOR

Dr. Alan R. Townsend is a scientist, author, speaker, and dean of the University of Montana's W. A. Franke College of Forestry and Conservation. He is a highly cited author of more than 140 scientific articles and has served in multiple prominent leadership roles. He was named an Aldo Leopold Leadership Fellow and a Google Science Communication Fellow and was one of six scientists chosen to be in the *Let Science Speak* documentary film series, which premiered at the Tribeca TV Festival in September of 2018. He lives in Montana with his family and two ridiculous dogs.